Adam Bayley, Andrew McAdam, Matthew Tristley
Iain Wilson and Steve Wiseman

REVISION PLUS

OCR Twenty First Century
GCSE Physics A
Revision and Classroom Companion

Ideas about Science

Introduction to Ideas about Science

The OCR Twenty First Century Physics specification aims to ensure that you develop an understanding of science itself – of how scientific knowledge is obtained, the kinds of evidence and reasoning behind it, its strengths and limitations, and how far we can rely on it.

These issues are explored through Ideas about Science, which are built into the specification content and summarised over the following pages.

The tables below give an overview of the Ideas about Science that can be assessed in each unit and provide examples of content which support them in this guide.

Unit A181 (Modules P1, P2 and P3)

Ideas about Science	Example of Supporting Content
Cause–effect explanations	Tectonic Plates (page 8)
Developing scientific explanations	Global Warming (page 18)
The scientific community	Radiation Protection (page 16)
Risk	Radiation Protection (page 16)
Making decisions about science and technology	Generating Electricity (page 22)

Unit A182 (Modules P4, P5 and P6)

Ideas about Science	Example of Supporting Content
Data: their importance and limitations	Nuclear Fusion (page 53)
Cause–effect explanations	Collisions (page 40)
Developing scientific explanations	Nuclear Fusion (page 53)
Risk	Uses of Radiation (page 56)
Making decisions about science and technology	Tracers in the Body (page 57)

Unit A183 (Module P7)

Ideas about Science	Example of Supporting Content
Data: their importance and limitations	Measuring Distance Using Brightness (page 72)
Cause–effect explanations	Pressure, Volume and Temperature (page 74)
Developing scientific explanations	The Hubble Constant (page 73)
The scientific community	The Curtis–Shapley debate (page 73)
Risk	Space-based Telescopes (page 80)
Making decisions about science and technology	Funding Developments in Science (page 80)

❶ Data: Their Importance and Limitations

Science is built on data. Physicists carry out experiments to collect and interpret data, seeing whether the data agree with their explanations. If the data do agree, then it means the current explanation is more likely to be correct. If not, then the explanation has to be changed.

Experiments aim to find out what the 'true' value of a quantity is. Quantities are affected by errors made when carrying out the experiment and random variation. This means that the measured value may be different to the true value. Physicists try to control all the factors that could cause this uncertainty.

Physicists always take repeat readings to try to make sure that they have accurately estimated the true value of a quantity. The mean is calculated and is the best estimate of what the true value of a quantity is. The more times an experiment is repeated, the greater the chance that a result near to the true value will fall within the mean.

The range, or spread, of data gives an indication of where the true value must lie. Sometimes a measurement will not be in the zone where the majority of readings fall. It may look as if the result (called an 'outlier') is wrong – however, it does not automatically mean that it is. The outlier has to be checked by repeating the measurement of that quantity. If the result cannot be checked, then it should still be used.

Here is an example of an outlier in a set of data:

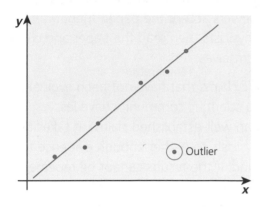

The spread of the data around the mean (the range) gives an idea of whether it really is different to the mean from another measurement. If the ranges for each mean do not overlap, then it is more likely that the two means are different. However, sometimes the ranges do overlap and there may be no significant difference between them.

The ranges also give an indication of reliability – a wide range makes it more difficult to say with certainty that the true value of a quantity has been measured. A small range suggests that the mean is closer to the true value.

If an outlier is discovered, you need to be able to defend your decision as to whether you keep it or discard it.

❷ Cause–effect Explanations

Science is based on the idea that a factor has an effect on an outcome. Physicists make predictions as to how the input variable will change the outcome variable. To make sure that only the input variable can affect the outcome, physicists try to control all the other variables that could potentially alter it. This is called 'fair testing'.

You need to be able to explain why it is necessary to control all the factors that might affect the outcome. This means suggesting how they could influence the outcome of the experiment.

A correlation is where there is an apparent link between a factor and an outcome. It may be that as the factor increases, the outcome increases as well. On the other hand, it may be that when the factor increases, the outcome decreases. For example, there is a correlation between temperature and the rate of rusting – the higher the temperature, the faster the rate of rusting.

Just because there is a correlation does not necessarily mean that the factor causes the outcome. Further experiments are needed to establish this. It could be that another factor causes the outcome or that both the original factor and outcome are caused by something else.

The following graph suggests a correlation between going to the opera regularly and living longer. It is far more likely that if you have the money to go to the opera, you can afford a better diet and health care. Going to the opera is not the true cause of the correlation.

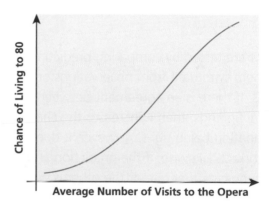

Sometimes the factor may alter the chance of an outcome occurring but does not guarantee it will lead to it. The statement 'the more time spent on a sun bed the greater the chance of developing skin cancer' is an example of this type of correlation, as some people will not develop skin cancer even if they do spend a lot of time on a sun bed.

To investigate claims that a factor increases the chance of an outcome, physicists have to study groups of people who either share as many factors as possible or are chosen randomly to try to ensure that all factors will present in people in the test group. The larger the experimental group, the more confident physicists can be about the conclusions made.

Ideas about Science

❸ Developing Scientific Explanations

Physicists devise hypotheses (predictions of what will happen in an experiment), along with an explanation (the scientific mechanism behind the hypotheses) and theories (that can be tested).

Explanations involve thinking creatively to work out why data have a particular pattern. Good scientific explanations account for most or all of the data already known. Sometimes they may explain a range of phenomena that were not previously thought to be linked. Explanations should enable predictions to be made about new situations or examples.

When deciding on which is the better of two explanations, you should be able to give reasons why.

Explanations are tested by comparing predictions based on them with data from observations or experiments. If there is an agreement between the experimental findings, then it increases the chance of the explanation being right. However, it does not prove it is correct. Likewise, if the prediction and observation indicate that one or the other is wrong, it decreases the confidence in the explanation on which the prediction is based.

❹ The Scientific Community

Once a physicist has carried out enough experiments to back up his/her claims, they have to be reported. This enables the scientific community to carefully check the claims, something which is required before they are accepted as scientific knowledge.

Physicists attend conferences where they share their findings and sound out new ideas and explanations. This can lead to physicists revisiting their work or developing links with other laboratories to improve it.

The next step is writing a formal scientific paper and submitting it to a journal in the relevant field. The paper is allocated to peer reviewers (experts in their field), who carefully check and evaluate the paper. If the peer reviewers accept the paper, then it is published. Physicists then read the paper and check the work themselves.

New scientific claims that have not been evaluated by the whole scientific community have less credibility than well-established claims. It takes time for other physicists to gather enough evidence that a theory is sound. If the results cannot be repeated or replicated by themselves or others, then physicists will be sceptical about the new claims.

If the explanations cannot be arrived at from the available data, then it is fair and reasonable for different physicists to come up with alternative explanations. These will be based on the background and experience of the physicists. It is through further experimentation that the best explanation will be chosen.

This means that the current explanation has the greatest support. New data are not enough to topple it. Only when the new data are sufficiently repeated and checked will the original explanation be changed.

❺ Risk

Everything we do (or not do) carries risk. Nothing is completely risk-free. New technologies and processes based on scientific advances often introduce new risks.

Risk is sometimes calculated by measuring the chance of something occurring in a large sample over a given period of time (calculated risk). This enables people to take informed decisions about whether the risk is worth taking. In order to decide, you have to balance the benefit (to individuals or groups) with the consequences of what could happen.

For example, deciding whether or not to have radiotherapy for the treatment of cancer involves weighing up the benefit (of being treated) against the risk (of side effects).

Risk which is associated with something that someone has chosen to do is easier to accept than risk which has been imposed on them.

> **HT** Perception of risk changes depending on our personal experience (perceived risk). Familiar risks (e.g. smoking) tend to be under-estimated, whilst unfamiliar risks (e.g. radiotherapy) and invisible or long-term risks (e.g. radiation) tend to be over-estimated.
>
> For example, many people under-estimate the risk of getting skin cancer from exposure to ultraviolet rays.

Governments and public bodies try to assess risk and create policy on what is and what is not acceptable. This can be controversial, especially when the people who benefit most are not the ones at risk.

6 Making Decisions about Science and Technology

Science has helped to create new technologies that have improved the world, benefiting millions of people. However, there can be unintended consequences of new technologies, even many decades after they were first introduced. These could be related to the impact on the environment or to the quality of life.

When introducing new technologies, the potential benefits must be weighed up against the risks.

Sometimes unintended consequences affecting the environment can be identified. By applying the scientific method (making hypotheses, explanations and carrying out experiments), physicists can devise new ways of putting right the impact.

Some areas of physics could have a high potential risk to individuals or groups if they go wrong or if they are abused. In these areas the Government ensures that regulations are in place.

The scientific approach covers anything where data can be collected and used to test a hypothesis. It cannot be used when evidence cannot be collected (e.g. it cannot test beliefs or values).

Just because something can be done does not mean that it should be done. Some areas of scientific research or the technologies resulting from them have ethical issues associated with them. This means that not all people will necessarily agree with it.

Ethical decisions have to be made, taking into account the views of everyone involved, whilst balancing the benefits and risks.

It is impossible to please everybody, so decisions are often made on the basis of which outcome will benefit most people. Within a culture there will also be some actions that are always right or wrong, no matter what the circumstances are.

Contents

Scientific discoveries in the solar system affect our understanding of the planet we live on and our place in the Universe. This module looks at:
- what is known about Earth
- how the Earth's continents have moved and the consequences
- what is known about stars and galaxies
- how scientists develop explanations about Earth and space
- how waves travel.

The Solar System

The **solar system** was formed over a very long period of time, about **5000 million years** ago.

1. The solar system started as **clouds of dust and gas**, which were pulled together by the **force of gravity** (see diagram below).
2. This created intense heat. Eventually, **nuclear fusion** began to take place and a star was born: **the Sun**.
3. The remaining dust and gas formed **smaller masses**, which were attracted to the Sun.

The smaller masses in our solar system are:
- **planets** – eight large masses that orbit (move around) the Sun
- **moons** – small masses that orbit the planets
- **asteroids** – small, rocky masses that orbit the Sun
- **comets** – small, icy masses that orbit the Sun
- **dwarf planets** – small spherical objects that have not cleared their orbits of other objects.

Planets, moons and asteroids all move in **elliptical** (slightly squashed circular) orbits. Comets move in highly elliptical orbits (see diagram below). It takes our planet, Earth, **one year** to make a complete orbit around the Sun.

The Sun

The Sun's **energy** (heat and light) comes from **nuclear fusion**. **Hydrogen** nuclei **fuse** (join) together to produce a nucleus with a larger mass, i.e. a new chemical element. During fusion, some of the energy trapped inside the hydrogen nuclei is released. All the **chemical elements** with a larger mass than helium were formed by nuclear fusion in **earlier stars**.

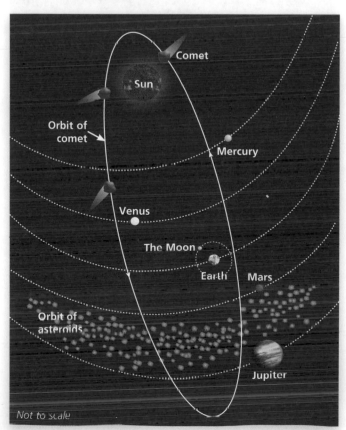

Not to scale

The Universe

At 5000 million years old, the Sun is only 500 million years older than the **Earth**. The **Universe** is much older than this: approximately **14 000 million years old** (almost three times older than the Sun).

The Sun is one of thousands of millions of stars in the Milky Way.

Not to scale

Our planet – the Earth,
12 800km diameter

Our star – the Sun, 100 times wider than Earth

Our Sun

The solar system is approximately
10 billion km across

Not to scale

Our galaxy – the Milky Way, 100 000 light-years
across, containing at least 100 billion stars

Our galaxy

The Universe – contains billions of galaxies,
with vast distances between them

The Speed of Light

Light travels at very **high but finite** (limited) **speeds**. This means that if the distance to an object is great enough, the time taken for light to get there can be **measured**.

The speed of light is **300 000km/s** in a vacuum (around one million times faster than sound). So, light from Earth takes just over one second to reach the Moon (approximately 384 400km away).

Light from the Sun takes eight minutes to reach the Earth. This means that when we look at the Sun, we are actually seeing what it looked like eight minutes ago.

Vast distances in space are measured in **light-years**. One light-year is the **distance light travels in one year** (approximately 9500 billion km). The nearest galaxy to the Milky Way is 2.2 million light-years away. This means that light from this galaxy has taken 2.2 million years to reach the Earth, and so we are seeing the galaxy as it was in the past.

Measuring Distances in Space

Astronomers work out the distances to different **stars** using two different methods:

1 Relative brightness

In general, the dimmer a star is, the further away it is. However, stars can vary in brightness so we can never be 100% certain.

2 Parallax

If you hold out a finger at arm's length and close each eye in turn, the finger appears to move. The closer the finger is to your face, the more it appears to move. Parallax uses this idea to work out distances.

As the Earth orbits the Sun, stars in the near distance appear to move against the background of very distant stars. The closer they are, the more they appear to move.

The position of a star is measured at six-monthly intervals. These measurements can then be used to calculate its distance from Earth. However, the further away the star is, the more difficult and less accurate the measurement is.

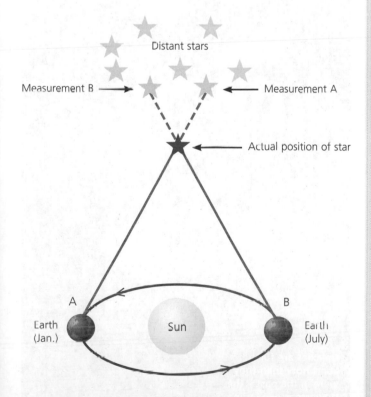

Distant Stars

Because the stars are so far away, everything we know about them is worked out from the **radiation** they produce: **visible light** and other types of radiation, including **ultraviolet** and **infrared**.

All our electric lights on Earth illuminate the night sky, so it is very difficult to see the stars sometimes. This is called **light pollution**. In 1990 the **Hubble Space Telescope** was launched. It orbits the Earth at a height of 600km, so it is not affected by light pollution or atmospheric conditions.

Other Galaxies

If a source of light is moving away from us, the **wavelengths** of the light are **longer** than they would be if the source was stationary.

The wavelengths of light from almost all **galaxies** are longer than scientists would expect. This means the **galaxies are moving away from us**.

HT In 1929 Edwin Hubble discovered that light from distant galaxies had even longer wavelengths. Therefore, they must be moving away from us faster. As a result, he developed **Hubble's Law**:

The speed at which galaxies are moving away from us is proportional to their distance from us (i.e. the faster a galaxy is moving, the further away it is).

If all the galaxies are moving away from one another, this must mean that space is **expanding** (getting bigger).

Redshift

If a wave source is moving away from or towards an observer, there will be a change in the observed wavelength and frequency.

If a source of light moves away from an observer, the wavelengths of the light in its spectrum are longer than if it was not moving. This is known as **redshift** because the wavelengths 'shift' towards the red end of the spectrum.

All distant galaxies appear to be 'redshifted', which means they are moving away from us.

The Beginning

When scientists trace the paths of galaxies, they all appear to be moving away from the same point.

There have been many theories about how the Universe began. The one that best explains this evidence is the **Big Bang** theory, which says that the Universe started with a **huge explosion 14 000 million years ago**.

The End

It is difficult to predict the fate of the Universe because it is very hard to measure the very large distances involved. It is also very difficult to study the motion of very distant objects.

The future depends on the amount of **mass** in the Universe. If there is not enough mass, the Universe will keep expanding. If there is too much mass, gravity will be strong enough to pull everything back together and the Universe will collapse with a big crunch. Measuring the amount of mass in the Universe is very difficult, so its ultimate fate is hard to predict.

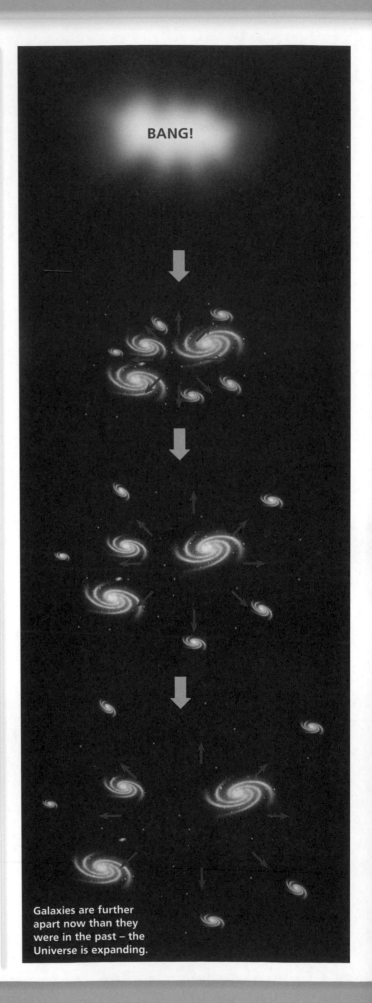

BANG!

Galaxies are further apart now than they were in the past – the Universe is expanding.

The Earth

People once thought that the Earth was only 6000 years old. There was no way of testing this theory, so people believed it for a long time.

We now know that **rocks** provide evidence of how the Earth has changed and clues as to its age.

Erosion – the Earth's surface is made up of **layers** of rock, one on top of the other, with the oldest at the bottom. The layers are made of compacted **sediment**, which is produced by weathering and **erosion**. Erosion changes the surface of the planet over long periods of time.

Craters – the surface of the Moon is covered with impact **craters** from collisions with meteors. However, the Earth, which is much larger, has had fewer meteor collisions (due to Earth's atmosphere), but craters have also been erased by erosion.

Mountain formation – if new **mountains** were not being formed, the whole Earth would have been worn down to sea level by erosion.

Fossils – plants and animals trapped in layers of sedimentary rock have formed **fossils**, providing evidence of how life on Earth has changed over millions of years.

Folding – some rocks look as if they have been **folded** like plasticine. This would require a big force to be applied over a long period of time – further evidence that the Earth is very old.

Radioactive dating – all rocks are **radioactive**, but the amount of radiation they emit **decreases** over time. Radioactive dating measures radiation levels to find out how old they are.

Scientists estimate that the Earth is around **4500 million years old** – it has to be older than its oldest rocks – and when it was first formed it was completely **molten** (hot liquid) and would have taken a very long time to cool down.

The oldest rocks that have been found on Earth are about **4000 million years old**.

The Structure of the Earth

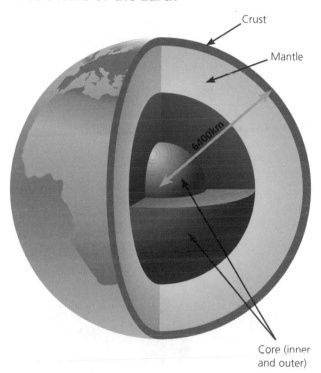

Crust

Mantle

6400km

Core (inner and outer)

Thin rocky crust:
- Its thickness varies between 10km and 100km.
- Oceanic crust lies beneath the oceans.
- Continental crust forms continents.

The mantle:
- Extends almost halfway to the centre.
- Has a higher density, and a different composition, than rock in the crust.
- Very hot, but under pressure.

The core:
- Made of nickel and iron.
- Over half of the Earth's radius; has a liquid outer part and a solid inner part.
- The decay of radioactive elements inside the Earth releases energy, which keeps the interior of the Earth hot.

Continental Drift

Alfred Wegener (1880–1930) was a meteorologist who put forward a theory called **continental drift**.

He saw that the continents all fitted together like a jigsaw, with the mountain ranges and sedimentary rock patterns matching up almost perfectly. There were also fossils of the same land animals on different continents. Wegener proposed that the different continents were once joined together, but had become separated and drifted apart.

How It Once Was

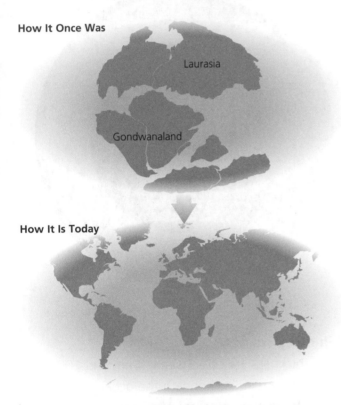

How It Is Today

Wegener also claimed that when two continents collided, they forced each other upwards to make mountains.

Geologists at the time did not accept Wegener's theory because:
- he was not a geologist and was therefore considered to be an outsider
- it was a big idea but he was not able to provide much evidence
- the evidence could be explained more simply by a land bridge connecting the continents that has now sunk or been eroded
- the movement of the continents was not detectable.

Wegener's evidence for continental drift could be summarised as:
- similar patterns of rocks, which contain fossils of the same plants and animals, e.g. the Mesosaurus
- closely matching coastlines.

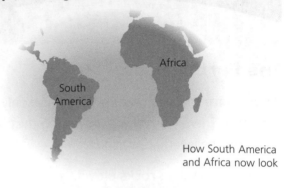

How South America and Africa now look

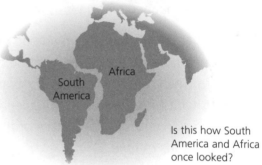

Is this how South America and Africa once looked?

Tectonic Plates

We now know that the Earth's crust is cracked into several large pieces called **tectonic plates.**

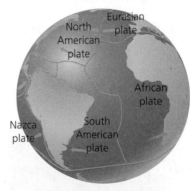

The plates float on the Earth's mantle because they are less dense. They can move apart, move towards each other or slide past each other. The lines where the plates meet are called **plate boundaries**. These are where **volcanoes**, **earthquakes** and **mountain building** normally occur.

Earthquakes that occur near coastlines or at sea can often result in a **tsunami** (similar to a tidal wave).

Tectonic Plate Movement

The movement of the tectonic plates can happen suddenly due to a build up in pressure and can sometimes have disastrous consequences, e.g. earthquakes and tsunamis. Tectonic plates can move in three ways:

1 Slide Past Each Other

When plates slide, huge stresses and strains build up in the crust which eventually have to be released in order for movement to occur. This 'release' of energy results in an earthquake. A classic example of this is the West Coast of North America (especially California).

2 Move Away from Each Other – Constructive Plate Boundaries

When plates move away from each other at an oceanic ridge, fractures occur. Molten rock rises to the surface, where it solidifies to form new ocean floor. This is known as seafloor spreading. Because new rock is being formed, these are called **constructive** plate boundaries.

3 Move Towards Each Other – Destructive Plate Boundaries

As plates are moving away from each other in some places, it follows that they must be moving towards each other in other places. When plates collide, one is forced under the other, so these are called **destructive** plate boundaries. Earthquakes and volcanoes are common on destructive plate boundaries.

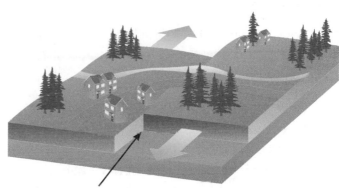

An earthquake will occur along the line where the two plates meet

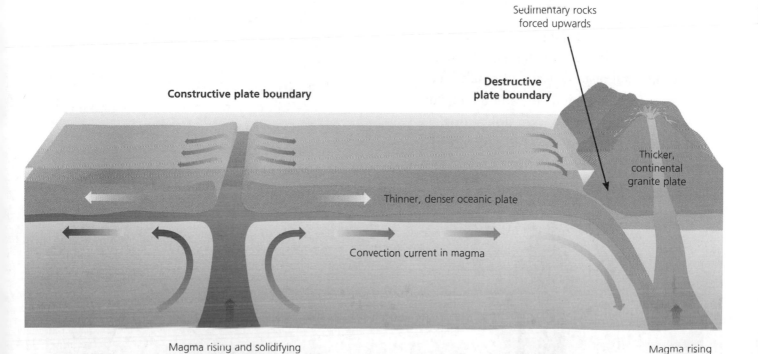

Sedimentary rocks forced upwards

Constructive plate boundary

Destructive plate boundary

Thicker, continental granite plate

Thinner, denser oceanic plate

Convection current in magma

Magma rising and solidifying to form new ocean floor (a few centimetres per year)

Magma rising up through continental crust

Seafloor Spreading

Just below the Earth's crust the mantle is fairly solid. Further down it is liquid and able to move. **Convection currents** in the mantle cause magma (molten rock) to rise to the surface. The force is strong enough to move the solid part of the mantle and the tectonic plates. When the magma reaches the surface, it hardens to form new areas of oceanic crust (seafloor), pushing the existing floor outwards.

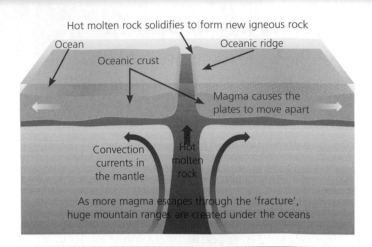

Hot molten rock solidifies to form new igneous rock

Ocean

Oceanic crust

Oceanic ridge

Magma causes the plates to move apart

Convection currents in the mantle

Hot molten rock

As more magma escapes through the 'fracture', huge mountain ranges are created under the oceans

Plate Tectonics

New oceanic crust is continuously forming at the crest of an oceanic ridge and old rock is gradually pushed further outwards.

The Earth has a **magnetic field** which changes polarity (reverses) every million years or so. Combined with the spreading of the seafloor, this produces stripes of rock of alternating polarity. Geologists can work out how quickly new crust is forming from the widths of the stripes. This occurs at **constructive plate boundaries**, where the plates are moving apart.

When an oceanic plate and a continental plate collide, the denser oceanic plate is forced under the continental plate. This is called **subduction**. The oceanic plate then melts, and the molten rock can rise upwards to form **volcanoes**. The boundaries where this occurs are called **destructive plate boundaries**.

Mountain ranges form along plate boundaries as sedimentary rock is forced upwards by the pressure created in a collision.

Earthquakes occur most frequently at plate boundaries when plates slide past each other or collide. Pressure builds up over many years due to the force of the plates pushing against each other. Eventually, the stored energy is released in a sudden upheaval of the crust and spreads outwards in waves from the epicentre.

Plate movement plays a crucial role in the **rock cycle**:

- Old rock is destroyed through subduction.
- Igneous rock is formed when magma reaches the surface.
- Plate collisions can produce very high temperatures and pressures, causing the rock to fold and changing sedimentary rock into metamorphic rock.

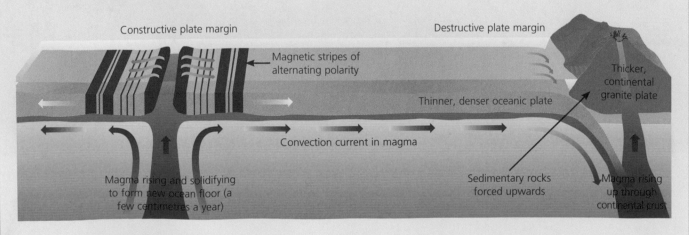

Constructive plate margin

Magnetic stripes of alternating polarity

Destructive plate margin

Thicker, continental granite plate

Thinner, denser oceanic plate

Convection current in magma

Magma rising and solidifying to form new ocean floor (a few centimetres a year)

Sedimentary rocks forced upwards

Magma rising up through continental crust

Evidence for the Structure of the Earth

Evidence for the layered structure of the Earth has been gained through the study of earthquakes. These are due to the fracture of large masses of rock inside the Earth. The energy that is released travels through the Earth as a series of shock waves called seismic waves, which are detected using seismographs.

There are two types of shock waves: P-waves and S-waves. Differences in the speed of P- and S-waves can be used to give evidence for the structure of the Earth.

P-waves

HT • Longitudinal waves (see page 12) where the ground is made to vibrate in the same direction as the shock wave is travelling, i.e. if the shock wave is travelling from left to right the ground also vibrates from left to right.

• Pass through solids and liquids.
• Faster than S-waves.
• Speed increases in denser material.

S-waves

HT • Transverse waves (see page 12) where the ground is made to vibrate at right angles to the direction the shock wave is travelling, i.e. if the shock wave is travelling from left to right the ground vibrates up and down.

• Pass through solids only.
• Slower than P-waves.
• Speed increases in denser material.

P-waves – building vibrates left to right

S-waves – building vibrates up and down

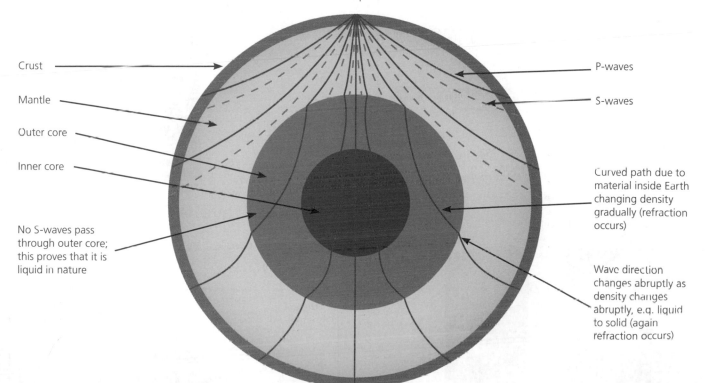

Earthquake

Crust

Mantle

Outer core

Inner core

No S-waves pass through outer core; this proves that it is liquid in nature

P-waves

S-waves

Curved path due to material inside Earth changing density gradually (refraction occurs)

Wave direction changes abruptly as density changes abruptly, e.g. liquid to solid (again refraction occurs)

No S-waves on the opposite side to the earthquake

Types of Waves

Waves are regular patterns of disturbance that transfer energy in the direction the wave travels without transferring matter. There are two types of wave – **longitudinal** and **transverse**.

All waves transfer energy from one point to another without transferring particles of matter. If we consider that each coil of the slinky spring in the following diagrams represents one particle, then we can show the movement of the particles in each type of wave.

Longitudinal Waves

Each particle moves backwards and forwards in the same plane as the direction of wave movement. Each particle simply vibrates to and fro about its normal position.

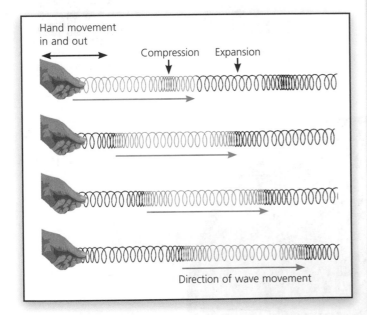

Sound travels as longitudinal waves.

Transverse Waves

Each particle moves up and down at right angles (90°) to the direction of wave movement. Each particle simply vibrates up and down about its normal position.

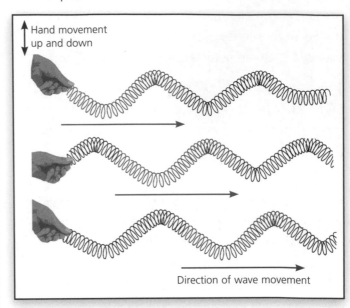

Light and water ripples both travel as transverse waves.

Distance Travelled by a Wave

The distance travelled by a wave can be worked out using the formula:

| Distance (metres, m) | = | Wave speed (metres per second, m/s) | × | Time (seconds, s) |

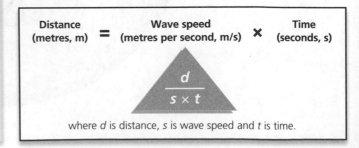

where d is distance, s is wave speed and t is time.

Wave Features

All waves have several important features:
- **Amplitude** – the maximum disturbance caused by a wave. It is measured by the distance from a crest or trough of the wave to the undisturbed position.
- **Wavelength** – the distance between corresponding points on two adjacent disturbances.
- **Frequency** – the number of waves produced, (or passing a particular point) in one second. Frequency is measured in hertz (Hz).

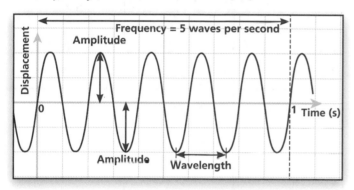

Wave Speed, Wavelength and Frequency

If a wave travels at a **constant speed** (i.e. its speed does not change), then:
- increasing its frequency will decrease its wavelength
- decreasing its frequency will increase its wavelength.

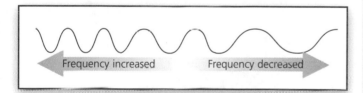

If a wave has a **constant frequency** (i.e. its frequency does not change), then:
- decreasing its wave speed will decrease its wavelength
- increasing its wave speed will increase its wavelength.

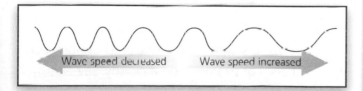

N.B. The speed of a wave is usually independent of its frequency and amplitude.

The Wave Equation

Wave speed, frequency and wavelength are related by the following formula:

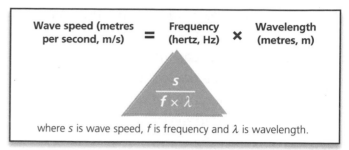

where s is wave speed, f is frequency and λ is wavelength.

For a constant wave speed, the wavelength is inversely proportional to the frequency.

Example

A tapped tuning fork of frequency 480Hz produces sound waves with a wavelength of 70cm. What is the speed of the wave?

Using our relationship:

Wave speed = Frequency × Wavelength

$\qquad$ = 480Hz × 0./m $\longleftarrow$ Wavelength must be in metres.

$\qquad$ = **336m/s**

HT The wave speed formula can be rearranged using the formula triangle to work out the frequency or wavelength.

Example

Radio 5 Live transmits on a frequency of 909 000Hz. If the speed of radio waves is 300 000 000m/s, on what wavelength does it transmit?

Rearrange the formula:

Wave speed = Frequency × Wavelength

Wavelength $= \dfrac{\text{Wave speed}}{\text{Frequency}}$

$\qquad = \dfrac{300\,000\,000\text{m/s}}{909\,000\text{Hz}} =$ **330m**

Radiation is all around us and, while there are some dangers and hazards, certain radiations are essential to life on this planet. This module looks at:

- the electromagnetic spectrum
- how radiation is transmitted
- ionising radiation and any harmful effects
- what happens to sunlight entering the Earth's atmosphere
- the carbon cycle and evidence of global warming
- how information is added to a wave.

The Electromagnetic Spectrum

The electromagnetic spectrum is a family of seven **radiations**, including visible light.

A beam of electromagnetic radiation contains 'packets' of energy called **photons**. Different radiations contain photons that carry different amounts of energy.

The intensity of a beam of radiation depends on the number of these **photons** it delivers every second. The intensity of the beam also depends upon the amount of energy carried by each photon.

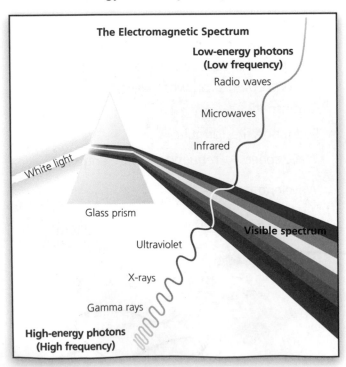

The Electromagnetic Spectrum

Low-energy photons (Low frequency)

Radio waves

Microwaves

Infrared

White light

Glass prism

Visible spectrum

Ultraviolet

X-rays

Gamma rays

High-energy photons (High frequency)

The higher the frequency of the electromagnetic radiation, the more energy is transferred by each photon.

Transmitting Radiation

All electromagnetic waves travel through space (a vacuum) at the same very high speed (300 000 km/s).

A general model of radiation describes how energy travels from a source that emits radiation to a detector that absorbs radiation.

Examples

Emitter	How Waves Travel	Detector
TV transmitter	Radio waves	TV aerial
Mobile phone mast	Microwaves	Mobile phones
The Sun	Light	The eye
Remote control	Infrared	Television
Some stars (e.g. supernova)	Gamma rays	Gamma-ray telescope
X-ray machine	X-rays	Photographic plate

On the journey from emitter to detector the radiation can be transmitted, reflected or absorbed by materials. For example, on a cloudy day energy from the Sun is absorbed and reflected by the clouds, and the amount of light received at the ground is less than it would be on a sunny day.

Intensity and Heat

The **intensity** (or measure of strength) of electromagnetic radiation is the **energy** arriving at a **square metre** of surface **per second**.

The intensity depends on the number of photons delivered per second and the amount of energy each individual packet contains, i.e. the photon energy.

The intensity of a beam of radiation decreases with distance, so the further away from a source you are, the lower the intensity.

> **HT** This decrease in intensity is due to three factors:
> - The photons spread out as they travel so the energy is more spread out.
> - Some of the photons are absorbed by particles in the substances they pass through.
> - Some of the photons are reflected and scattered by other particles.
>
> These effects combine to reduce the number of photons arriving per second at a detector, resulting in a lower measured intensity.

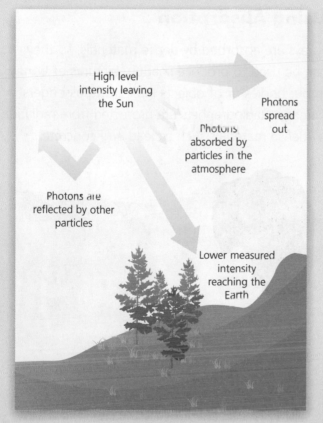

High level intensity leaving the Sun

Photons spread out

Photons absorbed by particles in the atmosphere

Photons are reflected by other particles

Lower measured intensity reaching the Earth

When a material absorbs radiation, it will heat up; the temperature increase depends on the intensity of the radiation.

The amount of heating also depends on the duration of exposure.

Some electromagnetic radiations (ultraviolet, X-rays, gamma rays) have enough energy to change atoms or molecules.

> **HT** These changes can initiate a chemical reaction.

Ionising Radiation

Some materials (radioactive materials) emit ionising gamma radiation all the time. Ionising radiation has photons with enough energy to remove an electron from an atom or molecule to form **ions**.

Ionising radiations are those with high enough photon energy to remove an electron from an atom or molecule. Ultraviolet radiation, X-rays and gamma rays are all examples of ionising radiation.

> **HT** Ions are very reactive and can easily take part in other chemical reactions.

Cell Damage

When living cells absorb radiation, damage can occur in different ways:
- The heating effect can cause damage.
- Ionising radiation, such as ultraviolet radiation, can damage cells, causing ageing of the skin.
- Ionising radiation can cause mutations in the nucleus of a cell, which can lead to cancer.
- Different amounts of exposure can cause different effects, e.g. high-intensity ionising radiation can kill cells, leading to radiation poisoning.

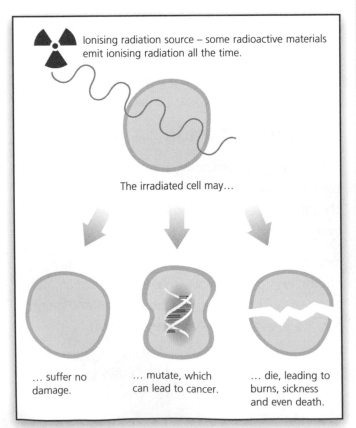

Ionising radiation source – some radioactive materials emit ionising radiation all the time.

The irradiated cell may…

… suffer no damage.

… mutate, which can lead to cancer.

… die, leading to burns, sickness and even death.

Radiation Protection

Microwaves are strongly absorbed by water molecules, which means microwaves can be used to heat objects containing water.

Microwave ovens have a metal case and a wire screen in the door – this reflects the microwaves and protects users by preventing too much radiation from escaping. The door screen also absorbs microwaves, protecting users from the radiation.

There may be a health risk from the low-intensity microwaves of mobile phone handsets and masts, but this is disputed. A study in 2005 found no link from short-term use, but other studies have found some correlation between mobile phone masts and health problems. Further studies are underway to look in more detail at mobile phone masts and the long-term effects of mobile phone use.

Other physical barriers are used to protect people from ionising radiation, e.g. sun-screens and clothing can absorb most of the ultraviolet radiation from the Sun, and this helps to prevent skin cancer.

Using Absorption

X-rays are absorbed by dense materials, so they can be used to produce shadow pictures of bones in our bodies or of objects in aircraft passengers' luggage. Radiographers are protected from radiation by dense materials such as lead and concrete.

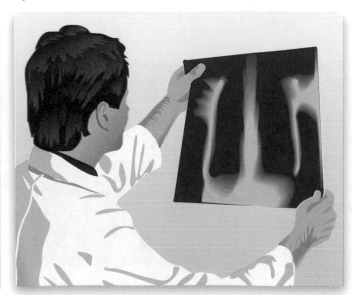

The Sun's Energy

The Sun (and all other objects) emits electromagnetic radiation with a principal frequency that increases with temperature. The Earth is surrounded by an atmosphere that allows some of the electromagnetic radiation emitted by the Sun to pass through.

This radiation warms the surface of the Earth when it is absorbed.

The Ozone Layer

The ozone layer is a thin layer of gas in the Earth's upper atmosphere. This layer of gas absorbs some of the ultraviolet radiation from the Sun before it can reach Earth.

Without the ozone layer, the amount of ultraviolet radiation reaching Earth would be very harmful to living organisms, especially animals, due to cell damage.

> (HT) The energy from the ultraviolet radiation causes chemical changes in the upper atmosphere when it is absorbed by the ozone layer, but these changes are reversible.

The Greenhouse Effect

The Earth emits electromagnetic radiation into space.

> (HT) This radiation has a lower principal frequency than that emitted by the Sun.

There are gases in the atmosphere that absorb or reflect some of this radiation. This keeps the Earth warmer than it would otherwise be and is known as the **greenhouse effect**.

Greenhouse Gases

Carbon dioxide is a **greenhouse gas** and it makes up a small amount of the Earth's atmosphere – about 0.035%. Other greenhouse gases include water vapour and very small amounts of methane.

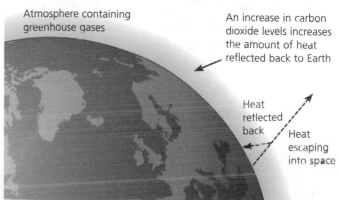

Atmosphere containing greenhouse gases

An increase in carbon dioxide levels increases the amount of heat reflected back to Earth

Heat reflected back

Heat escaping into space

The Carbon Cycle

The carbon cycle is an example of a balanced system.

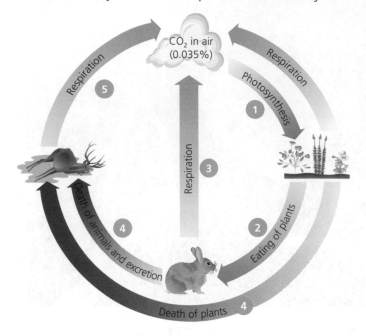

1. Carbon dioxide is removed from the atmosphere by green plants to produce glucose by photosynthesis. Some is returned to the atmosphere by the plants during respiration.
2. The carbon obtained by photosynthesis is used to make carbohydrates, fats and proteins in plants. When the plants are eaten by animals, this carbon becomes carbohydrates, fats and proteins in animals.
3. Animals respire, releasing carbon dioxide into the atmosphere.
4. When plants and animals die, other animals and microorganisms feed on their bodies, causing them to break down. Excretion also releases carbon.
5. As the detritus feeders and microorganisms eat the dead plants and animals, they respire, releasing carbon dioxide into the atmosphere.

The carbon cycle can be used to explain several points:

- The amount of **carbon dioxide** in the atmosphere had remained roughly constant for thousands of years because it was constantly being recycled by plants and animals.
- The importance of **decomposers**, which are microorganisms that break down dead material and release carbon dioxide back into the atmosphere.

- The amount of carbon dioxide in the atmosphere has been steadily increasing over the last 200 years, largely due to human activity such as burning fossil fuels and deforestation.
- **Fossil fuels** contain carbon that was removed from the atmosphere millions of years ago and has been 'locked up' ever since. Burning fossil fuels for energy releases this carbon into the atmosphere.
- **Burning forests** (**deforestation**) to clear land not only releases the carbon they contain but also reduces the number of plants removing carbon dioxide from the atmosphere.

Global Warming

The increase of greenhouse gases in the Earth's atmosphere, especially carbon dioxide, means that the amount of absorbed radiation from the Sun also increases. This increases the temperature on Earth, an effect known as global warming. As the Earth becomes hotter, there are some potential results:

- **Climate change** – it may become impossible to grow some food crops in certain areas.
- **Extreme weather conditions** (e.g. floods, droughts, hurricanes).

> HT These are caused by increased convection and larger amounts of water vapour in the hotter atmosphere.

- **Rising sea levels** – the melting ice caps and higher ocean temperatures may cause sea levels to rise, which could cause flooding of low-lying land. Some Pacific islands have already been abandoned.

HT Causes of Global Warming

Climatologists collect data about how the Earth's temperature has changed over the years. The data collected is used with climate models to look for patterns. These computer models show that one of the main factors causing global warming is the rise in atmospheric carbon dioxide and other greenhouse gases, providing evidence that human activities are causing global warming.

Uses of Electromagnetic Waves in Communication

Different electromagnetic waves have different frequencies. This affects their properties and the effect that other materials have on them.

They can be used for different purposes, depending on how much they are reflected, absorbed or transmitted by different materials.

Electromagnetic Waves	Properties and Uses
Radio Waves	• Radio waves are used for transmitting radio and television programmes because they are not strongly absorbed by the Earth's atmosphere. They can travel long distances through the atmosphere and through space. • Radio telescopes are used in astronomy to pick up radio waves from stars.
Microwaves	• Microwaves are used to transmit mobile phone signals because they are not strongly absorbed by the atmosphere. • They are reflected well by metals so satellite dishes are made of metal and shaped to reflect the signal onto the receiver.
Light and Infrared Radiation	• Light and infrared radiation will travel huge distances down optical fibres without the signal becoming significantly weaker. This makes them very useful for carrying information, e.g. in computer networks and telephone conversations.

Transmitting Information

For communication purposes, information can be superimposed onto an electromagnetic carrier wave, to create a signal.

For a wave to carry a signal, it must be modulated. This involves changing the carrier wave to create a variation that matches the variation in the information that is being transmitted. It is this pattern of variation that carries the information.

The pattern of variation is decoded by the receiver to reproduce the original information, e.g. sound from a radio wave.

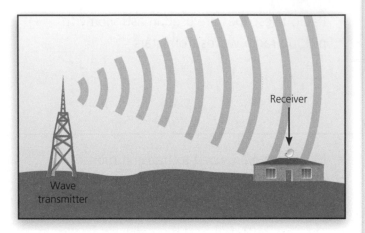

Analogue Signals

An **analogue** signal can vary continuously, so its amplitude and/or frequency can take any value. A sound wave is an example of an analogue signal. Some radio stations are transmitted as analogue signals.

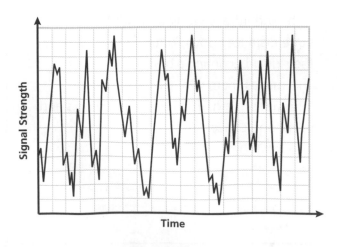

Digital Signals

Information, including sound and images, can also be transmitted digitally. The signal is not sent as a continuously varying transmission. A **digital** signal can only take one of a small number of fixed (discrete) values – usually two. For transmitting information digitally, the digital code is made up of just two symbols, '1' and '0'.

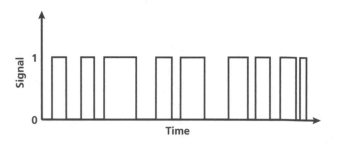

The electromagnetic carrier wave is switched off and on to create short bursts of waves. A receiver decodes the pulses to re-create the original information. Mobile phone signals, digital radio and digital TV are all transmitted in this way.

Benefits of Digital

Both digital and analogue signals become weaker (their amplitude becomes smaller) as they travel. The transmitted signals therefore have to be amplified at selected intervals to make them stronger.

During transmission the signals can also pick up random variations, called **noise**, which reduce the quality of the signal.

With digital signals it is easier to remove the noise, and so recover the original signal. This gives digital signals an important advantage over analogue signals. Information transmitted digitally can travel long distances and is received at a higher quality than analogue signals because any interference can be removed.

Another advantage of using a digital signal is that the information can be stored and processed by computers. Generally, the more information stored (or the more **bytes**, **B**), the higher the quality of the sound or image.

Signal Quality

Analogue signals can have many different values, so it is hard to distinguish between noise and the original signal. This means that noise cannot be completely removed and, when the signal is amplified, any noise that has been picked up is also amplified.

With analogue signals you are often left with a noisy signal. With music this can involve a 'hissing' sound.

Digital signals, which have two states, on (1) or off (0), can still be recognised despite any noise that is picked up. Therefore, it is easy to remove the noise and clean up the signal, restoring the on / off pattern. So, when it is amplified, the quality of the digital signal is retained.

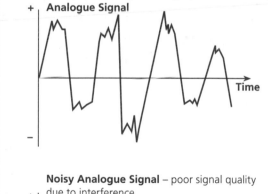

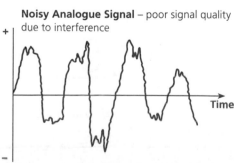

Noisy Analogue Signal – poor signal quality due to interference

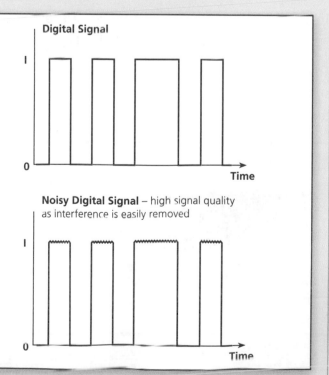

Digital Signal

Noisy Digital Signal – high signal quality as interference is easily removed

Module P3 (Sustainable Energy)

Energy is generated from a variety of sources and the electricity is used in the home. This module looks at:

- how electricity is generated
- renewable and non-renewable energy sources
- generators and power
- the cost of electricity
- energy efficiency.

Electricity

Electricity is called a **secondary energy source** because it is generated from another energy source, e.g. coal, nuclear, wind, etc.

During the generation process some energy is always lost to the surroundings. This makes electricity less efficient than when compared with using the primary resource directly.

However, the convenience of electricity makes it very useful. It can be easily transmitted over long distances and used in a variety of ways.

Generating Electricity

To generate electricity, fuel (either fossil fuel or nuclear) is used to release energy as heat.

The heat is used to boil water which produces steam, and the steam is then used to drive the turbines that power the generators.

The electricity produced in the generators is sent to a transformer and then to the National Grid. The voltage is then reduced to a safer level of 230V, after which we can access it in our homes.

In a **fossil fuel** power station the fuel is burned to release the chemical energy it contains as heat. As they are burning carbon fuels, the power stations also produce carbon dioxide, a greenhouse gas.

In a **nuclear** power station the energy is released due to changes in the nucleus of radioactive substances. Nuclear power stations do not produce carbon dioxide, but they do produce **radioactive waste**.

Nuclear waste emits ionising radiation, and this can cause a number of health-related issues:

- **Irradiation** means exposure to radiation. This can happen naturally through background radiation from sources such as the Earth or space. It could also happen by exposure during medical treatments such as X-rays. In such cases the cells may become damaged, but the person does not become radioactive. Increased exposure may eventually lead to cancer and death.
- **Contamination** involves a radioactive material being placed inside a person. This can be far more damaging than irradiation, yet it is often used in medical treatments where the risk is considered worth the benefit, such as tumour suppression.

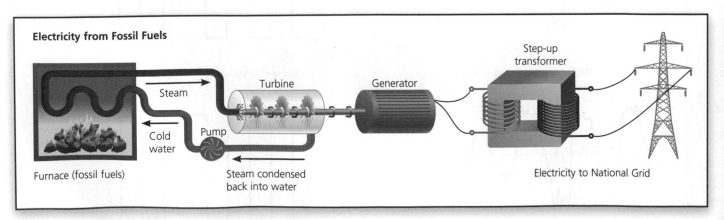

Electricity from Fossil Fuels

Steam

Cold water

Pump

Furnace (fossil fuels)

Steam condensed back into water

Turbine

Generator

Step-up transformer

Electricity to National Grid

Non-renewable Energy Sources

Coal, oil and gas are energy sources that are formed over millions of years from the remains of plants and animals. They are called **fossil fuels** and are responsible for most of the energy that we use. However, because they cannot be replaced within a lifetime, they will eventually run out. They are therefore called **non-renewable** energy sources.

Coal   Oil   Gas

Nuclear fuels such as uranium and plutonium are also non-renewable. Nuclear fission is the splitting of a nucleus that generates thousands of times more heat energy than burning the same mass of fossil fuel.

However, nuclear fuel is not burned like coal, oil or gas to release energy and is not classed as a fossil fuel.

Renewable Energy Sources

As the demand for electricity continually increases, other sources of energy are needed. Renewable energy sources are those that will not run out because they are continually being replaced. Many of them are caused by the Sun or Moon. The gravitational pull of the Moon creates tides, and the Sun causes:

- evaporation, which results in rain and flowing water
- convection currents, which result in winds, that in turn create waves.

Generating Electricity from Renewable Energy Sources

Renewable energy sources can be used to drive turbines or generators directly. In other words, no fuel needs to be burned to produce heat.

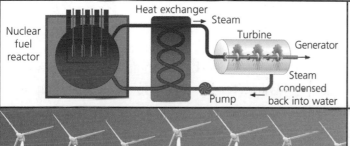	**Nuclear Fuel (Non-renewable)** A nuclear reactor is used to generate heat by nuclear fission. A heat exchanger is used to transfer the heat energy from the reactor to the water, which turns to steam and drives the turbines.
	**Wind Turbines (Renewable)** Wind can be used to drive huge turbines which, in turn, drive generators. Wind turbines are positioned in exposed places where there is a lot of wind, such as the tops of hills or offshore.
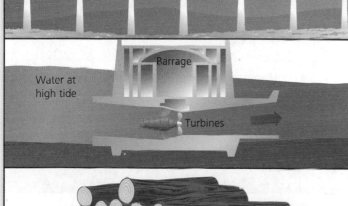	**Tidal Barrage (Renewable)** As the tide comes in, water flows freely through a valve in the barrage. This water then becomes trapped. At low tide, the water is released from behind the barrage through a gap which has a turbine in it. This drives a generator.
	**Wood (Renewable)** Although burned for energy, wood is not a fossil fuel, nor is it non-renewable. It is classed as a **renewable** energy source since trees can be grown relatively quickly to replace those that are burned to provide energy for heating.

P3 | Sustainable Energy

Comparing Non-renewable Sources of Energy

The energy sources below provide most of the electricity we need in this country through power stations.

Source	Advantages	Disadvantages
Gas	• Enough natural gas left for the short to medium term. • Can be found as easily as oil. • No sulfur dioxide (SO_2) is produced. • Gas-fired power stations are flexible in meeting demand and have a quicker start-up time than nuclear, coal and oil-fired reactors.	• Burning produces carbon dioxide (CO_2), although it produces less than coal and oil per unit of energy. (CO_2 contributes to global warming and climate change.) • Expensive pipelines and networks are often required to transport it to the point of use.
Coal	• Relatively cheap and easy to obtain. • Coal-fired power stations are flexible in meeting demand and have a quicker start-up time than their nuclear equivalents. • Estimates suggest that there may be over a century's worth of coal left.	• Burning produces CO_2 and SO_2. • Produces more CO_2 per unit of energy than oil or gas does. • SO_2 causes acid rain unless the sulfur is removed before burning or the SO_2 is removed from the waste gases. Both of these add to the cost of electricity.
Oil	• Enough oil left for the short to medium term. • Relatively easy to find, though the price is variable. • Oil-fired power stations are flexible in meeting demand and have a quicker start-up time than both nuclear-powered and coal-fired reactors.	• Burning produces CO_2 and SO_2. • Produces more CO_2 than gas per unit of energy. • Often carried between continents on tankers leading to the risk of spillage and pollution.
Nuclear	• Cost of fuel is relatively low. • Nuclear power stations are flexible in meeting demand. • No CO_2 or SO_2 produced.	• Although there is very little escape of radioactive material in normal use, radioactive waste can stay dangerously radioactive for thousands of years and safe storage is expensive. • Building and decommissioning is costly. • Longest comparative start-up time.

Summary of Non-renewable Resources

Advantages	Disadvantages
• Produce huge amounts of energy. • Reliable. • Flexible in meeting demand. • Do not take up much space (relatively).	• Pollute the environment. • Cause global warming and acid rain (fossil fuels only). • Will eventually run out. • Fuels often have to be transported over long distances.

Comparing Renewable Sources of Energy

The energy sources below use modern technology to provide a clean, safe alternative source of energy.

Source	Advantages	Disadvantages
Wind	• No fuel and little maintenance required. • No pollutant gases produced. • Once built, wind turbines provide 'free' energy when the wind is blowing. • Can be built offshore.	• Need a lot to produce a sizeable amount of electricity, which means noise and visual pollution. • Electricity output depends on the wind. • Not very flexible in meeting demand. • Capital outlay can be high to build turbines.
Tidal and Waves	• No fuel required. • No pollutant gases produced. • Once built, installations provide 'free' energy. • Barrage water can be released when demand for electricity is high.	• Tidal barrages across estuaries are unsightly, a hazard to shipping, and destroy the habitats of wading birds, etc. • Daily variations of tides and waves affect output. • High initial capital outlay to build barrages.
Hydro-electric	• No fuel required unless storing energy to meet future demand. • Fast start-up time to meet growing demand. • Produces a lot of clean, reliable electricity. • No pollutant gases produced. • Water can be pumped back up to the reservoir when demand for electricity is low, e.g. in the night.	• Location is critical and often involves damming upland valleys, which means flooding farms, forests and natural habitats. • To achieve a net output (aside from pumping) there must be adequate rainfall in the region where the reservoir is. • Very high initial capital outlay (though worth the investment in the end).
Solar	• Ideal for producing electricity in remote locations. • Excellent energy source for small amounts. • Produces free, clean electricity. • No pollutant gases produced.	• Dependent on the intensity of light. • High cost per unit of electricity produced, compared to all other sources except non-rechargeable batteries.
Bio-fuels	• Contain no sulfur (responsible for acid rain). • Can use many readily available waste materials. • Could be considered carbon neutral.	• Have lower energy output than traditional fuels. • Could lead to competition of land use between fuel and food.
Geo-thermal	• Minimal fuel costs. • Long life span and varying size of power output.	• High initial capital costs. • Possible environmental damage from harmful gases escaping from deep within the Earth.

Summary of Renewable Sources

Advantages	Disadvantages
• No fuel costs during operation. • Generally no chemical pollution. • Often low maintenance.	• Some produce small amounts of electricity. • Can be unreliable. • High initial capital outlay for most.

HT You may be asked to interpret and evaluate information about different energy sources for generating electricity, considering power output and lifetime.

The Electric Generator

Mains electricity is produced by generators. Generators use the principle of electromagnetic induction to generate electricity by rotating a magnet inside a coil.

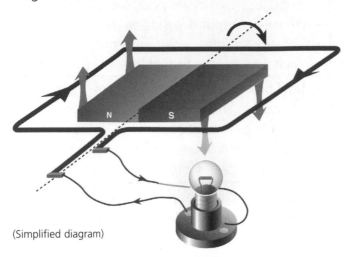

(Simplified diagram)

Power

When charge flows through a component, energy is transferred to the component. Power, measured in watts (W), is a measure of how much energy is transferred every second.

🔵 Power is therefore **the rate of energy transfer**.

Electrical power can be calculated using the following equation:

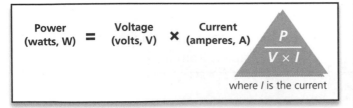

| Power (watts, W) | = | Voltage (volts, V) | × | Current (amperes, A) |

$$\frac{P}{V \times I}$$

where *I* is the current

Example

A hairdryer has a current of 3A and a potential difference of 230V. What is the power of the hairdryer?

Power = Voltage × Current
= 230V × 3A
= 690W

🔵 The power formula can be rearranged using the formula triangle to work out the potential difference or current.

Example
What current is needed to power a 6W light bulb with a potential difference of 3V?

$$\text{Current} = \frac{\text{Power}}{\text{Voltage}}$$

$$= \frac{6W}{3V}$$

= 2A

Energy

Energy is measured in joules (J). A joule is a very small amount of energy, so a domestic electricity meter measures the energy transfer in a much larger unit, the kilowatt hour (kWh).

The amount of energy transferred for either joules or kilowatt hours can be calculated by the following equation:

| Energy transferred (joules, J) | = | Power (watts, W) | × | Time (seconds, s) |
| Energy transferred (kilowatt hours, kWh) | = | Power (kilowatts, kW) | × | Time (hours, h) |

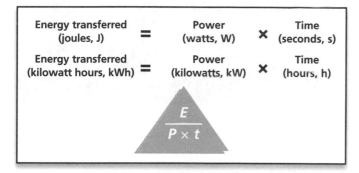

$$\frac{E}{P \times t}$$

Examples

1. A 30W light bulb is switched on for 45 seconds. What is the energy transferred in joules?

   Energy transferred = Power × Time
   = 30W × 45s
   = **1350J**

2. A 2000W electric hot plate is switched on for 90 minutes. What is the energy transferred in kilowatt hours?

   Energy transferred = 2kW × 1.5h ← Power in kW and time in hours
   = **3kWh**

HT The energy transfer formula can be rearranged using the formula triangle to work out power or time.

Example
A hairdryer is switched on for six minutes and the total energy transferred is 0.2kWh. What is the power rating of the hairdryer?

$$\text{Power} = \frac{\text{Energy transferred}}{\text{Time}} = \frac{0.2\text{kWh}}{0.1\text{h}} = \textbf{2kW}$$

Cost of Electricity

The cost of the electrical energy used can be calculated if the power, time and cost per kilowatt hour are known.

The formula for the cost of energy is as follows:

| Total cost | = | Number of kWh | × | Cost per kWh |

Example
A 2000W electric fire is switched on for 30 minutes. How much does it cost if electricity is 8p per kWh?

Energy transferred = 2kW × 0.5h
= 1kWh

Cost = 1kWh × 8p
= **8 pence**

Efficiency of Appliances

The greater the proportion of energy that is usefully transferred, the more efficient we say an appliance is. Efficiency can be calculated using the following formula:

$$\text{Efficiency (\%)} = \frac{\text{Energy usefully transferred}}{\text{Total energy supplied}} \times 100\%$$

Examples

Electrical Appliance	Energy In	Useful Energy Out	Efficiency
Light bulb	100 joules/s	Light: 20 joules/s	$\frac{20}{100} \times 100\%$ = **20%** or **0.2**
Kettle	2000 joules/s	Heat (in water): 1800 joules/s	$\frac{1800}{2000} \times 100\%$ = **90%** or **0.9**
Electric motor	500 joules/s	Kinetic: 300 joules/s	$\frac{300}{500} \times 100\%$ = **60%** or **0.6**
Television	200 joules/s	Light: 20 joules/s Sound: 30 joules/s	$\frac{50}{200} \times 100\%$ = **25%** or **0.25**

Sustainable Energy

Losing Energy

Energy is lost at every stage of the process of electricity generation. **Sankey diagrams** can be used to show the generation and distribution of electricity, including the efficiency of energy transfers.

The diagram below shows that of the energy put into the power station, almost half is lost to the surroundings (mostly as heat) before the electricity even reaches the home. When energy passes through any electrical component or device, energy will either be transferred to the device or to the environment.

Further energy is lost during energy transfers in the home when the electricity is used.

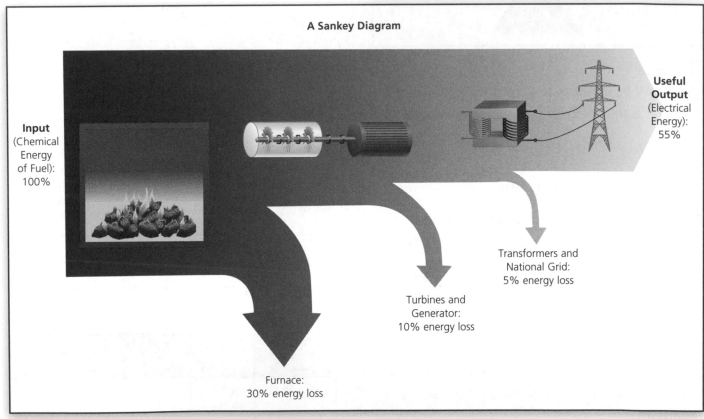

A Sankey Diagram

Input
(Chemical Energy of Fuel): 100%

Useful Output (Electrical Energy): 55%

Transformers and National Grid: 5% energy loss

Turbines and Generator: 10% energy loss

Furnace: 30% energy loss

The National Grid

The greater the power supplied by a generator, the more of the primary fuel (coal, oil, etc.) it uses every second. Depending on the type of power station, voltages of between 1000 and 25 000V can be produced.

Due to the inefficiency of electrical energy transfer through cables, the National Grid uses extremely high voltages through its network of pylons. Over half a million volts is often used, meaning a low current is needed and less energy is lost due to heat. The voltages are then reduced to 230V by **transformers** in local substations before safely entering the home.

The Domestic Home

Energy saving in the home can be done in a variety of ways. These include loft insulation, double glazing, draught proofing, lagging the hot water tank and the use of efficient light bulbs.

When considering which methods to employ, the homeowner needs to weigh up the economic effectiveness of any changes.

For example, if you install energy-efficient, double-glazed windows and loft insulation, then when it's time to replace your boiler and heating system, you may be able to manage with a smaller one that costs less. This is because the windows and walls will retain the heated air inside better than a home without efficient windows and insulation.

The National Picture

When it comes to saving energy, choices are being made on a national scale as well as domestically. Reducing a country's carbon emissions is quickly becoming a global issue with international agreements such as the Kyoto Protocol.

> **HT** The need to reduce energy usage is important, but so is the need to maintain a secure supply of energy. This is done by using a wide mix of energy sources across the renewable and non-renewable range.

The Workplace

As well as energy loss through the physical infrastructure of a building, a large amount of energy can be saved by following a few simple steps in the workplace.

Turning off computers at the end of the working day, turning down an office radiator by just one degree Celsius or designing spaces to use more natural light can save money and help to reduce the business's carbon footprint.

Ways to Save Energy

In the Home	In the Workplace	National Context
• More efficient appliances, e.g. a condensing boiler could save £190 per year	• Cleaning air conditioner filters – can save 5% of the energy used in running the system	• Replacing old houses with new efficient ones
• Double glazing – possible savings of £130 per year	• Using low-energy light bulbs	• Increased use of public transport
• Loft insulation – possible savings of £145 per year	• Roof insulation / cavity wall insulation in modern buildings	• More efficient trains and buses
• Cavity wall insulation – possible savings of £110 per year	• Use of efficient modern, low energy machinery	• Encourage more widespread recycling
• Draught-proof rooms – possible savings of £25 per year	• Use of modern, efficient vehicles for transport of goods	• Encourage car sharing and fewer journeys

Exam Practice Questions

1 Studying the solar system around us has helped scientists have a much greater understanding of the Earth and its existence in the Universe.

(a) Place the following in order from smallest to largest:

A The Earth   **B** The Sun   **C** The Moon

D The solar system   **E** The Universe   **F** The galaxy   **[5]**

Smallest					Largest

(b) Approximately how old do scientists believe our solar system to be? **[1]**

(c) What is the speed of light? **[1]**

(d) What is a light-year? **[1]**

2 The electromagnetic spectrum has many uses, both good and bad.

(a) Complete the table below to show the electromagnetic spectrum in order of low-energy photons to high-energy photons. Choose words from this list:

Ultraviolet   Radio waves   Gamma rays   Infrared

Low-energy photons		Microwaves		Visible spectrum		X-rays		High-energy photons

[3]

(b) When living cells absorb radiation, damage can occur in a number of different ways. List some of the harmful effects of sunbathing. **[3]**

(c) Which parts of the electromagnetic spectrum are ionising and what does ionising mean? **[4]**

3 Electricity can be generated using a number of different energy resources, some of which are controversial.

(a) What is a **renewable** energy resource? **[1]**

(b) What is a **non-renewable** energy resource? **[1]**

(c) Which of the following energy resources are renewable? Put ticks (✓) in the boxes next to the three correct answers. **[3]**

☐ Coal ☐ Wind ☐ Wood ☐ Natural gas ☐ Oil ☐ Tidal water

(d) The burning of fossil fuels in power stations can cause a number of environmental issues. Explain how burning fossil fuels can lead to climate change. **[6]**

The quality of written communication will be assessed in your answer to this question.

(e) Calculate the amount of energy transferred by a 2000W kettle, when heating up water to make a cup of tea, in one minute. Use the following equation:

Energy transferred (J) = Power (W) ✗ Time (s)

[2]

4 What is **redshift** and what does it tell us about the Universe? **[2]**

5 The Earth's crust has been constantly changing over millions of years, leaving a variety of evidence for scientists to study.

(a) Use the theory of plate tectonics to explain how earthquakes are produced. **[4]**

(b) Explain why we have magnetic stripes of alternating polarity on either side of the Mid-Atlantic Ridge. **[6]**

The quality of written communication will be assessed in your answer to this question.

(c) By how much does the seafloor spread each year at the Mid Atlantic Ridge? **[1]**

Module P4 (Explaining Motion)

An understanding of the rules that govern forces and motion has led to the development of numerous modern technologies, from aeroplanes and spacecraft to sports equipment and theme park rides. This module looks at:

- how we can describe motion
- what forces are
- what the connection is between forces and motion
- how we can describe motion in terms of energy changes.

Speed and Velocity

Speed tells you how far an object will travel in a certain time. However, it does not tell you the direction of travel.

Velocity tells you an object's speed *and* gives an indication of its direction of travel.

Example

A lorry is travelling along a straight road. The speed of the lorry is 15m/s (metres per second).

In the morning, travelling in one direction (from the warehouse to the supermarket), the velocity is +15m/s. At night, travelling in the opposite direction (from the supermarket back to the warehouse), the velocity is -15m/s.

It does not matter which direction is called positive or negative as long as opposite directions have opposite signs.

This idea is also used when describing distance; changes in distance in one direction are described as positive, and in the opposite direction they are described as negative.

Calculating Speed

To calculate the speed of an object we need to know two things about its motion:

- The distance it has travelled.
- The time it took to travel that distance.

Speed is calculated using the following formula:

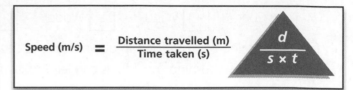

$$\text{Speed (m/s)} = \frac{\text{Distance travelled (m)}}{\text{Time taken (s)}}$$

The unit for speed is metres per second (m/s), so the distance should be in metres and the time in seconds.

A speed of 4m/s means that the object travels 4 metres every second. At this speed, in 2 seconds, the object would travel a total of 8 metres.

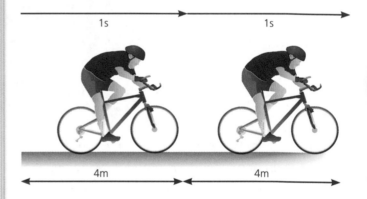

Speed can also be calculated in kilometres per hour, km/h.

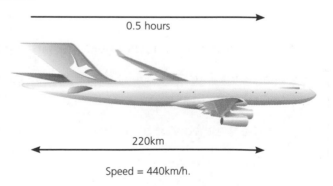

Speed = 440km/h.

The formula calculates an average speed over the total distance travelled, even if the speed of an object is not constant.

Example

An object travels 10 metres in 5 seconds. Use the formula to calculate its average speed.

$$\text{Speed} = \frac{\text{Distance travelled}}{\text{Time taken}} = \frac{10m}{5s} = \textbf{2m/s}$$

However, during those 5 seconds the object could have been moving at different speeds.

For example, the object could have moved at 5m/s for 1 second, stopped for 3 seconds, and then moved at 5m/s for 1 second. The object would still have an average speed of 2m/s, because it took 5 seconds to travel 10 metres.

The speed of an object at a particular point in time is called the **instantaneous speed**.

Distance–Time Graphs

A **distance–time graph** shows how the distance travelled by an object changes with time.

The slope or gradient of a distance–time graph is a measure of the speed of the object. The steeper the slope, the greater the speed.

The graph below shows a stationary person who is 10m from point 0.

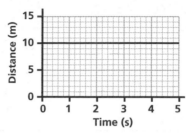

The graph below shows a person moving at a constant speed of 2m/s.

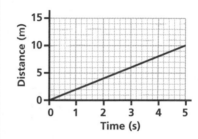

The graph below shows a person moving at a greater constant speed of 3m/s.

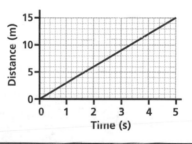

HT Distance–time graphs can also be drawn as **displacement–time graphs**, where the displacement of an object is its net distance from its starting point together with an indication of direction.

Example

The displacement–time graph opposite shows how the displacement of a football kicked against a wall changes with time. The ball is kicked, it hits the wall, and then rolls back between the person's legs before coming to a stop.

What is the distance covered by the ball and the displacement of the ball at the following times: 5s, 11s, 22s?

A Displacement–Time Graph for a Ball Kicked Against a Wall

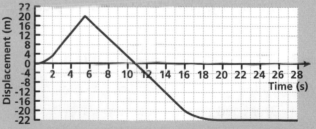

Time	Distance Covered	Displacement
5s	18m	18m
11s	20m + 20m = 40m	20m – 20m = 0
22s	20m + 20m + 22m = 62m	20m – 20m – 22m = -22m

The gradient of a displacement–time graph is the velocity

HT Calculating Speed

The speed of an object can be calculated by working out the gradient of a distance–time graph. The steeper the gradient, the faster the speed. All we have to do is take any point on the gradient and read off the distance travelled at this point, and the time taken to get there. Then use the formula:

$$\text{Speed} = \frac{\text{Distance travelled}}{\text{Time taken}}$$

Example
The graph below shows how the distance travelled by an object changes with time.

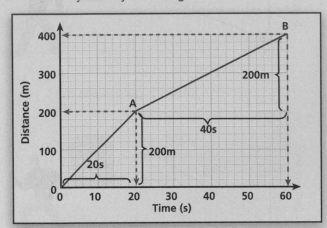

Find the speed of the object **(a)** from 0 to A, and **(b)** from A to B.

(a) From the graph, the distance from 0 to A is 200m and the time taken to travel this distance is 20s. Using the formula:

$$\text{Speed from 0 to A} = \frac{200m}{20s} = \textbf{10m/s}$$

(b) From the graph, the distance between A and B is 200m and the time taken to travel this distance is 60 – 20 = 40s. Using the formula:

$$\text{Speed from A to B} = \frac{200m}{40s} = \textbf{5m/s}$$

In this example, for the section A to B you need to find the *difference* in time and distance between points A and B. Do not just read the values from the axes. Using the values from the axes for point B will give you the *average* time for the whole journey from 0 to B, in this case 400m ÷ 60s = 6.7m/s.

The example above illustrates how to calculate speed from a graph, but remember that this only works when looking at straight-line sections.

Curvy Distance–Time Graphs

The slope or gradient of a distance–time graph represents how quickly an object is moving; the steeper the slope, the faster its speed. When the gradient is changing, the object's speed is changing.

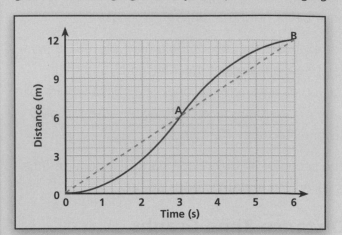

In section 0 to A the line is curved. This tells us that the object is speeding up because the gradient is increasing.

Between A and B the line curves the other way. The gradient is decreasing so the object must be slowing down.

Because the graph is curved it is difficult to work out the instantaneous speed at a particular point, but we can work out the average speed. The average speed is simply the total distance divided by the total time; it is shown by the dotted line.

$$\begin{aligned}\text{Speed} &= \frac{\text{Distance travelled}}{\text{Time taken}} \\ &= \frac{12m}{6s} \\ &= \textbf{2m/s}\end{aligned}$$

Where the gradient is steeper than the dotted line, the object is travelling faster than the average speed. Where the gradient is less steep than the dotted line, the object is travelling slower than the average speed.

Acceleration

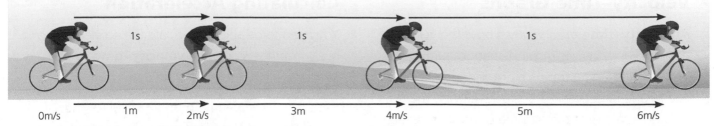

0m/s   1m   2m/s   3m   4m/s   5m   6m/s

The **acceleration** of an object is the rate at which its velocity changes. In other words, it is a measure of how quickly an object speeds up or slows down.

The cyclist above increases his velocity by 2 metres per second every second. So, we can say that his acceleration is 2m/s² (2 metres per second, per second). To work out the acceleration of any moving object you need to know two things:

- The change in velocity.
- The time taken for this change in velocity.

You can then calculate the acceleration of the object using the following formula:

$$\text{Acceleration (m/s}^2\text{)} \text{ (or deceleration)} = \frac{\text{Change in velocity (m/s)}}{\text{Time taken for change (s)}}$$

where *v* is the final velocity and *u* is the starting velocity

$$\frac{(v - u)}{a \times t}$$

There are two important points to be aware of:

1. The cyclist above increases his velocity by the *same amount* every second; the *actual* distance travelled each second increases.

2. Deceleration is simply a negative acceleration, i.e. it describes an object that is slowing down.

Example

A cyclist accelerates uniformly from rest and reaches a velocity of 10m/s after 5s, then decelerates uniformly and comes to a halt in a further 10s. Calculate his acceleration **(a)** in the first 5s, and **(b)** as he is slowing down in the next 10s.

(a) $\text{Acceleration} = \frac{\text{Change in velocity}}{\text{Time taken}}$

$= \frac{10 - 0}{5} = \textbf{2m/s}^2$

(b) $\text{Acceleration} = \frac{\text{Change in velocity}}{\text{Time taken}}$

$= \frac{0 - 10}{10} = \textbf{-1m/s}^2$

The minus sign shows this is a deceleration, so the deceleration is 1m/s².

Speed–Time Graphs

The slope of a **speed–time graph** represents the acceleration of the object; the steeper the slope, the greater the acceleration.

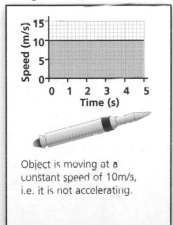

Object is moving at a constant speed of 10m/s, i.e. it is not accelerating.

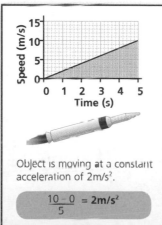

Object is moving at a constant acceleration of 2m/s².

$$\frac{10 - 0}{5} = 2\text{m/s}^2$$

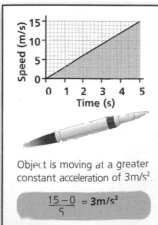

Object is moving at a greater constant acceleration of 3m/s².

$$\frac{15 - 0}{5} = 3\text{m/s}^2$$

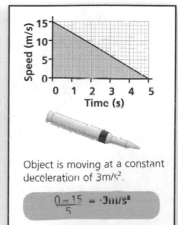

Object is moving at a constant deceleration of 3m/s².

$$\frac{0 - 15}{5} = -3\text{m/s}^2$$

Velocity–Time Graphs

A **velocity–time** graph shows how the velocity at which an object is moving changes with time. Velocity has a direction, so if moving in a straight line in one direction is a positive velocity, then moving in a straight line in the opposite direction will be a negative velocity.

The graph below is the velocity–time graph for someone swimming lengths in a swimming pool.

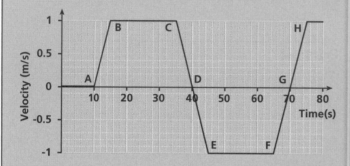

From 0 to A the swimmer is stationary – the velocity is zero. This is shown by a horizontal line along the time axis at zero velocity.

From A to B the swimmer is accelerating up the pool, and reaches a velocity of +1m/s after 15s.

Section B to C is a horizontal line at +1m/s. This shows that the swimmer is moving up the pool with constant velocity.

Section C to D shows the swimmer decelerating as they reach the end of the pool, turning round at D, where the velocity is zero once again.

Section D to E shows the swimmer accelerating to reach a velocity of -1m/s. The swimmer is now swimming at a constant velocity, but in the opposite direction, down the pool (section E to F).

Section F to G shows the swimmer decelerating to 0m/s at G, where they turn round. In section G to H the swimmer is once again accelerating up the pool, to reach a velocity of +1m/s.

Calculating Acceleration

You need to be able to calculate the acceleration of a body from the slope of a **velocity–time graph**.

Example

Here is a velocity–time graph. Calculate the acceleration of the three parts of the journey using the following formula:

$$\text{Acceleration} = \frac{\text{Change in velocity}}{\text{Time taken}}$$

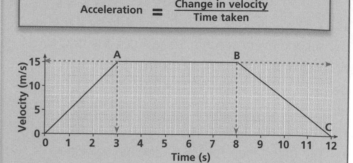

0 to A

From the graph, the time taken to reach a velocity of 15m/s is 3s.

Acceleration from 0 to A = $\frac{15\text{m/s}}{3\text{s}}$ = **5m/s²**

A to B

From A to B the line on the graph is horizontal, so the acceleration is zero.

Acceleration from A to B = $\frac{0\text{m/s}}{5\text{s}}$ = **0m/s²**

B to C

From B to C the velocity decreases from 15m/s to 0m/s in a time of 4s.

> The minus sign shows the object is slowing down, so this is a deceleration.

Acceleration from B to C = $\frac{-15\text{m/s}}{4\text{s}}$ = **-3.75m/s²**

So, the object accelerated at 5m/s² for 3 seconds, travelled at a constant speed of 15m/s for 5 seconds before decelerating at a rate of 3.75m/s² for 4 seconds.

Forces

A force occurs when two objects interact with each other. Whenever one object exerts a force on another, it always experiences an equal and opposite force in return. The forces in an **interaction pair** are equal in size and opposite in direction.

Examples

- **Gravity** (**weight**) – two masses are attracted to each other, e.g. we are attracted to the Earth and the Earth is attracted to us with an **equal** and **opposite** force. We do not notice the attraction of the Earth to us because it has such a big mass that our force has very little effect on it.

- If two people are standing on skateboards and one pushes the other, both skateboards will move away from each other.

- The rocket's engines push gas backwards (action) and the gas pushes the rocket forwards (reaction), thrusting it through the atmosphere. A jet engine works in a similar way.

Reaction: rocket goes up

Action: gas rushes down

Friction and Reaction

Some forces, such as friction and reaction (of a surface), only occur as a response to another force.

A force occurs when an object is resting on a surface. The object is being pulled down to the surface by gravity and the surface pushes up with an equal force called the **reaction of the surface**.

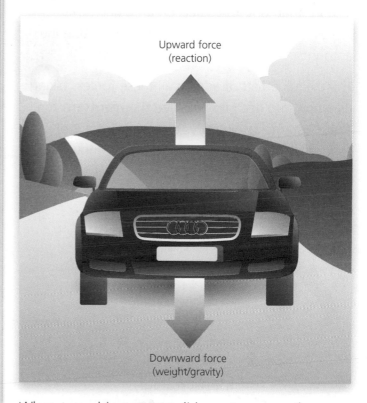

Upward force (reaction)

Downward force (weight/gravity)

When two objects try to slide past one another, both objects experience a force that tries to stop them moving. This interaction is called **friction**.

A moving object experiences friction. However, an object does not have to be moving to experience friction. For example, a car parked on a slope is trying to roll down the hill, but there is enough friction from the brakes (hopefully) to stop it.

Forces and Motion

Arrows are used when drawing diagrams of forces. The size and direction of the arrow represents the size of the force and the direction that it is acting in. Force arrows are always drawn with the tail of the arrow touching the object, even if the force is pushing the object.

If more than one force acts on an object, the forces need to be added, taking the direction into account.

The overall effect of adding all these forces, taking direction into account, is called the **resultant** force. The diagrams below show the overall effect of various push and friction forces on the movement of a trolley.

Trolley is Moving Forward

10N — Friction force
15N — Push force

Resultant force = ➔ 5N

Push force > Friction force, so accelerating

Trolley is Stationary

15N — Friction force
15N — Push force

Resultant force = 0

Push force = Friction force, so stationary

Trolley is Moving Forward

30N — Friction force
30N — Push force

Resultant force = 0N

Push force = Friction force, so constant velocity

Speeding Up and Slowing Down

A person walking, cars and bicycles have a **driving force**. In the case of a car, the driving force is produced by the engine. The person's muscles provide the driving force for someone walking or cycling.

However, there is also a **counter force**, caused by friction and air resistance, which has the effect of making the vehicle slow down or stop.

If the driving force is bigger than the counter force the vehicle accelerates.

100N   500N

Counter force 100N ⟵— Car —➔ Driving force 500N

Car speeds up

If the driving force and counter force are equal, the vehicle travels at a constant speed in a straight line.

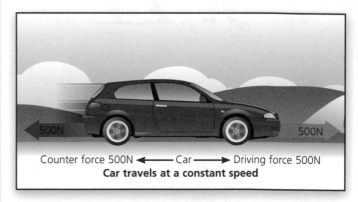

500N   500N

Counter force 500N ⟵— Car —➔ Driving force 500N

Car travels at a constant speed

If the counter force is bigger than the driving force, the vehicle decelerates.

1000N   500N

Counter force 1000N ⟵— Car —➔ Driving force 500N

Car slows down

Terminal Velocity

Falling objects experience two forces:

- the downward force of weight, W (↓) which always stays the same
- the upward force of air resistance, R, or drag (↑).

When a skydiver jumps out of an aeroplane, the speed of his descent can be considered in two separate parts – before and after the parachute opens:

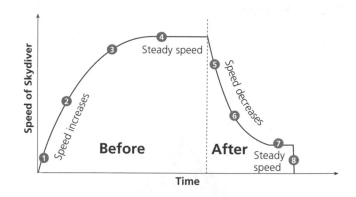

Before the Parachute Opens

When the skydiver jumps, he initially accelerates due to the force of gravity (see ❶). Gravity is a force of attraction that acts between objects that have mass, e.g. the skydiver and the Earth. The weight of an object is the force exerted on it by gravity. It is measured in newtons (N).

However, as the skydiver falls he experiences the frictional force of air resistance (R) in the opposite direction. But this is not as great as W so he continues to accelerate (see ❷).

As his speed increases, so does the air resistance acting on him (see ❸), until eventually R is equal to W (see ❹). This means that the resultant force acting on him is now zero and his falling speed becomes constant. This speed is called the **terminal velocity**.

After the Parachute Opens

When the parachute is opened, unbalanced forces act again because the upward force of R is now greatly increased and is bigger than W (see ❺). This decreases his speed, and as his speed decreases, so does R (see ❻).

Eventually R decreases until it is equal to W (see ❼). The forces acting are once again balanced, and for the second time he falls at a steady speed, slower than before though, i.e. at a new terminal velocity, until he lands on the ground (at ❽).

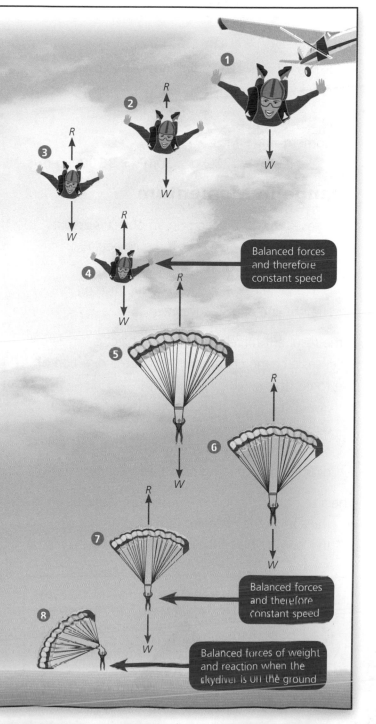

Balanced forces and therefore constant speed

Balanced forces and therefore constant speed

Balanced forces of weight and reaction when the skydiver is on the ground

Momentum

Momentum is a measure of the motion of an object. The momentum of an object is calculated using the following formula:

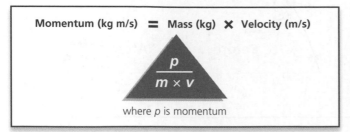

Momentum (kg m/s) = Mass (kg) × Velocity (m/s)

$$\frac{p}{m \times v}$$

where p is momentum

Example

A car has a mass of 1200kg. It is travelling at a velocity of 30m/s. Calculate its momentum.

Using the formula:

Momentum = Mass × Velocity

= 1200kg × 30m/s = **36 000kg m/s**

Change in Momentum

If the resultant force acting on an object is zero its momentum will not change. If the object:

- is stationary, it will remain stationary
- is already moving, it will continue moving in a straight line at a steady speed.

If the resultant force acting on an object is not zero, it causes a change in momentum in the direction of the force. This could:

- give a stationary object momentum (i.e. make it move)
- increase or decrease the speed, or change the direction, of a moving object, i.e. change its velocity.

The size of the change in momentum depends on:

- the size of the resultant force
- the length of time the force is acting on the object.

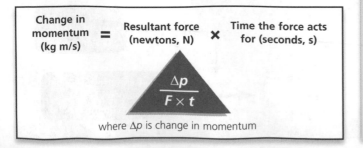

Change in momentum (kg m/s) = Resultant force (newtons, N) × Time the force acts for (seconds, s)

$$\frac{\Delta p}{F \times t}$$

where Δp is change in momentum

Collisions

If a car is involved in a collision it comes to a sudden stop, i.e. it undergoes a change in momentum. If this change in momentum is spread out over a longer period of time, the average resultant force acting on the car will be smaller.

Example

A car with a mass of 1000kg, travelling at 10m/s, has a momentum of 10 000kg m/s. If the car is involved in a collision and comes to a sudden stop, it would experience a change in momentum of 10 000kg m/s.

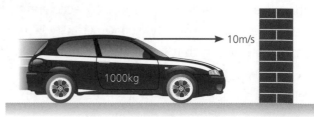
10m/s
1000kg

0m/s
1000kg

This sudden change in momentum will affect not only the car, but its passengers. If the car stops in a very short time, then the occupants of the car will experience a large force that can lead to serious injuries.

If the car stops in 0.05 seconds the force acting on it would be 200 000N, but if the car stops in 0.10 seconds the force would be 100 000N.

All car safety devices, such as seat-belts, crumple zones and air bags, are designed to reduce the force of the impact on the human body. They do this by increasing the time of the impact.

For example, a crumple zone is an area designed to 'crumple' on impact. This helps to increase the time during which the car changes momentum, i.e. instead of coming to an immediate halt there will be a longer time during which the momentum is reduced. This means that the force exerted on the people inside the car will be reduced, resulting in fewer injuries.

Kinetic Energy

A moving object has **kinetic** energy. The amount of kinetic energy an object has depends on:
- the mass of the object
- the velocity of the object.

The greater the mass and velocity of an object, the more kinetic energy it has. Therefore, it takes a longer time to stop a heavy, fast-moving object, than a light, slow-moving object.

Car brakes convert the kinetic energy into heat energy. As with all energy transfers, the total amount of energy remains the same (energy is conserved).

Kinetic energy is calculated using the following formula:

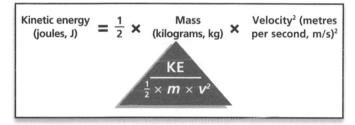

Kinetic energy (joules, J) $= \frac{1}{2} \times$ Mass (kilograms, kg) $\times$ Velocity^2 (metres per second, m/s)^2

$$\frac{KE}{\frac{1}{2} \times m \times v^2}$$

Example

A bicycle and rider have a mass of 100kg and are moving at a velocity of 8m/s. How much kinetic energy do they have?

8m/s

100kg

Using the formula:

$$\text{Kinetic energy} = \frac{1}{2} \times \text{Mass} \times \text{Velocity}^2$$
$$= \frac{1}{2} \times 100\text{kg} \times (8\text{m/s})^2$$
$$= \frac{1}{2} \times 100 \times 64$$
$$= \mathbf{3200J}$$

You need to be able to discuss the transformation of kinetic energy to other forms of energy in particular situations.

Example: Space Shuttles

When a space shuttle returns to Earth it has a lot of kinetic energy, i.e. it has a large mass and is travelling fast. As it enters the Earth's atmosphere, the shuttle encounters frictional forces and the kinetic energy is transformed into heat energy, which slows it down.

The shuttle can reach extremely high temperatures because of the heat energy produced, which causes a risk of fire and explosion. Scientists have developed special heat shields to try to protect the body of space shuttles (and the astronauts in them) from this intense heat.

Gravitational Potential Energy

An object lifted above the ground gains potential energy (PE), often called **gravitational potential energy (GPE)**. The additional height gives it the potential to do work when it falls, e.g. a diver standing on a diving board has gravitational potential energy. It is calculated by the following formula:

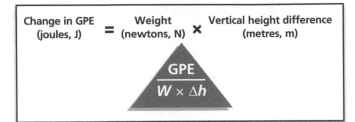

Change in GPE (joules, J) = Weight (newtons, N) × Vertical height difference (metres, m)

$$\frac{GPE}{W \times \Delta h}$$

If an object is dropped, its gravitational potential energy decreases and is converted into kinetic energy.

Example

On Earth, an object of mass 2kg weighs 20N. How much kinetic energy does the object gain if it is dropped from a height of 5m?

N.B. Remember, to find the GPE you use the weight (20N), not the mass.

Change in GPE = Weight × Height difference
= 20N × 5m = **100J**

If the object loses 100J of gravitational potential energy when it falls, it must have gained 100J of kinetic energy.

HT To work out the velocity of a falling object you need to use the mass. In the example above, we know that the object has gained 100J of kinetic energy.

Kinetic energy = $\frac{1}{2}$ × Mass × Velocity^2

$100 = \frac{1}{2} \times 2 \times velocity^2$

$100 = 1 \times velocity^2$

$100 = velocity^2$

$velocity = \sqrt{100}$

= **10m/s**

Work and Energy

When a force moves an object, work is done on the object resulting in the transfer of energy.

When work is done *by* an object it loses energy, and when work is done *on* an object it gains energy according to the relationship:

Amount of energy transferred (joules, J) = Work done (joules, J)

When a force acting on an object causes it to accelerate, work is done on the object to increase its kinetic energy. A falling object has work done on it by the force of gravity and this causes it to accelerate.

If we ignore the effects of air resistance and friction, the increase in kinetic energy will be equal to the amount of work done. However, in reality some of the energy will be dissipated (lost) as heat, and the increase in kinetic energy will, therefore, be less than the work done.

When a car is travelling at a constant speed the driving force from the engine is balanced by the counter force. All of the work being done by the engine is used to overcome friction and there is no increase in kinetic energy.

Work done is calculated by the following formula:

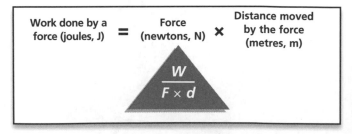

Work done by a force (joules, J) = Force (newtons, N) × Distance moved by the force (metres, m)

$$\frac{W}{F \times d}$$

Example

A car needs to be pushed with a force of 100N to overcome friction. If the car is pushed for 5m, calculate the work done.

Work done = Force × Distance moved
= 100N × 5m
= **500J**

Module P5 (Electric Circuits)

When electrical charges move, they create a current. Different types of current have different uses. This module looks at:

- static electricity
- electric currents
- electromagnetic induction
- electrical generators
- the use of transformers.

Static Electricity

Some insulating materials can become electrically charged when they are rubbed against each other. The electrical charge then stays on the material, i.e. it does not move (the charge is 'static').

Charge builds up when electrons (which have a negative charge) are 'rubbed off' one material onto another. The material receiving the electrons becomes negatively charged and the one giving up electrons becomes positively charged.

If a Perspex rod is rubbed with a cloth, the rod loses electrons to become positively charged. The cloth gains electrons to become negatively charged.

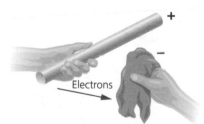

Perspex rod rubbed with cloth

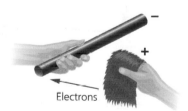

Ebonite rod rubbed with fur

If an ebonite rod is rubbed with fur, the ebonite gains electrons to become negatively charged. The fur loses electrons to become positively charged.

Repulsion and Attraction

When two charged materials are brought together, they exert a force on each other so they are attracted or repelled. Two materials with the same type of charge repel each other; two materials with different charges attract each other.

If a positively charged Perspex rod is moved near to another positively charged Perspex rod suspended on a string, the suspended rod will be repelled. We would get the same result with two negatively charged ebonite rods.

If a negatively charged ebonite rod is moved near to a positively charged suspended Perspex rod, the suspended Perspex rod will be attracted. We would get the same result if the rods were the other way round.

Example

When cars are spray painted, a panel of the car is positively charged and the paint is negatively charged. The paint particles repel each other, but are attracted to the positively charged panel. This causes the paint to be applied evenly.

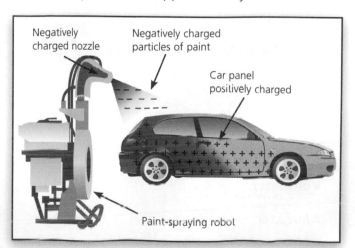

Electric Circuits

An **electric current** is a **flow of charge**. It is measured in **amperes** (amps).

In an electric circuit the components and wires are full of charges that are free to move. When a circuit is made, the battery causes these charges to move in a continuous loop. The charges are not used up.

In metal **conductors** there are lots of charges free to move. **Insulators**, on the other hand, have few charges that are free to move. Remember, metals contain free electrons in their structure. The movement of these electrons creates the flow of charge (electric current).

Direction of current

Circuit Symbols

You need to know the following standard symbols:

Cell	—⊣⊢—
Battery of cells	—⊣│⊢— or —⊣⁻⊣⊢—
Power supply	—o o—
Filament lamp	—⊗—
Switch (open)	—⟋ o—
Switch (closed)	—o o—
Light dependent resistor (LDR)	
Fixed resistor	—▭—
Variable resistor	
Thermistor	
Voltmeter	—(V)—
Ammeter	—(A)—

Current Size

The amount of current flowing in a circuit depends on the resistance of the components in the circuit and the **potential difference** (p.d.) across them. Potential difference tells us the energy given to the charge and is another name for **voltage**. It is a measure of the 'push' of the battery on the charges in the circuit.

For example, a 12 volt battery will transfer 12 joules of energy to every unit of charge. A potential difference of 3 volts across a bulb means that the bulb is transferring 3 joules of energy from every unit of charge; this energy is transferred as heat and light.

The greater the potential difference (or voltage) across a component, the greater the current that flows through the component. Two cells together provide a bigger potential difference across a lamp than one cell. This makes a bigger current flow.

Current is measured using an ammeter and voltage (or potential difference) is measured using a voltmeter (see page 46).

Resistance and Current

Components such as resistors, lamps and motors resist the **flow** of **charge** through them, i.e. they have **resistance**. Work is done by the power supply and energy is transferred to the component.

The greater the resistance of a component or components, the smaller the current that flows for a particular voltage, or the greater the voltage needed to maintain a particular current.

Even the connecting wires in the circuit have some resistance, but it is such a small amount that it is usually ignored.

Two lamps together in a circuit with one cell have a certain resistance. Including another cell in the circuit provides a greater potential difference, so a larger current flows.

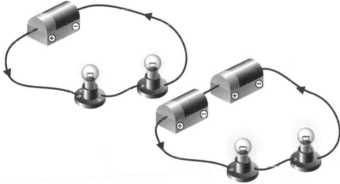

Adding resistors in series increases the resistance because the battery has to push charges through all of the resistors.

Adding resistors in parallel reduces the total resistance and increases the total current because this provides more paths for the charges to flow along.

When an electric current flows through a component it causes the component to heat up. In a filament lamp this heating effect is large enough to make the filament in the lamp glow.

> **HT** As the current flows, the moving charges collide with the vibrating ions in the wire giving them energy; this increase in energy causes the component to become hot.

Resistance, which is measured in ohms, Ω, is a measure of how hard it is to get a current through a component at a particular potential difference or voltage. Resistance can be calculated by the following formula:

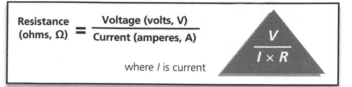

$$\text{Resistance (ohms, } \Omega) = \frac{\text{Voltage (volts, V)}}{\text{Current (amperes, A)}}$$

where *I* is current

Example

The circuit above has a current of 3 amps and a voltage of 6V. What is the resistance?

$$\text{Resistance} = \frac{\text{Voltage}}{\text{Current}} = \frac{6V}{3A} = \mathbf{2\Omega}$$

> **HT** The resistance formula can be rearranged using the formula triangle to work out the voltage or current.
>
> **Example**
> Calculate the reading on the voltmeter in this circuit if the bulb has a resistance of 15 ohms.
>
>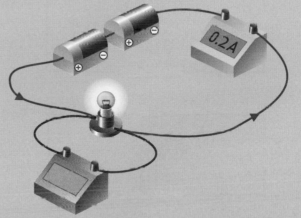
>
> Rearranging the formula:
> Voltage = Current × Resistance
> $$= 0.2A \times 15\Omega = \mathbf{3V}$$

Current–Voltage Graphs

As long as the temperature of a resistor stays constant, the current through the resistor is directly proportional to the voltage across the resistor, regardless of which direction the current is flowing, i.e. if one doubles, the other also doubles.

This means that a graph showing current through the component and voltage across the component will be a straight line through 0. (If there is no voltage, then there isn't anything to push the current around.)

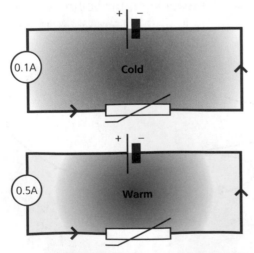

Thermistors and Light Dependent Resistors (LDRs)

The resistance of some materials depends on environmental conditions.

The resistance of a **thermistor** depends on its temperature. Its resistance decreases as the temperature increases; this allows more current to flow.

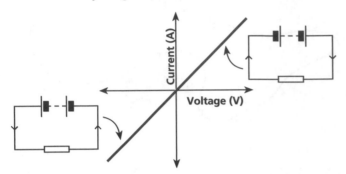

The resistance of a **light dependent resistor** (LDR) depends on light intensity. Its resistance decreases as the amount of light falling on it increases; this allows more current to flow.

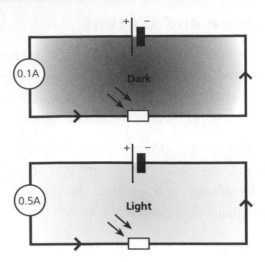

As the resistance of an LDR and a thermistor can change, this will result in a change in the potential difference for all the other components in the circuit.

Potential Difference and Current

The potential difference across a component in a circuit is measured in volts (V) using a voltmeter connected in parallel across the component.

The current flowing through a component in a circuit is measured in amperes (A), using an ammeter connected in series.

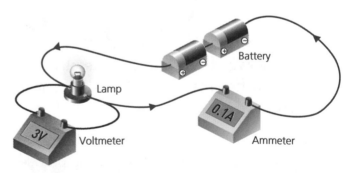

When batteries are added in series, the total potential difference is the sum of all the individual potential differences.

HT When batteries are added in parallel, the total potential difference and current through the circuit remains the same, but each battery supplies less current.

This sharing of the load makes the batteries last longer.

Series Circuits

In a series circuit, all components are connected one after another in one loop going from one terminal of the battery to the other.

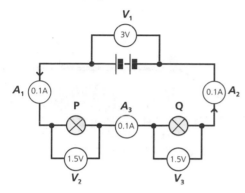

- The current flowing through each component is the same, i.e. $A_1 = A_2 = A_3$.
- The potential difference across the components adds up to the potential difference across the battery, i.e. $V_1 = V_2 + V_3$.

> **HT** The total energy transferred to each unit charge by the battery must equal the total amount of energy transferred from the charge by the component because energy cannot be created or destroyed.

- If another battery was added to the above circuit, it would have the effect of increasing the voltage and current.
- The potential difference is largest across components with the greatest resistance.

> **HT** More energy is transferred from the charge flowing through a greater resistance because it takes more energy to push the current through the resistor.
>
> If a circuit has several identical components, they will have the same potential difference across them. If the components are different, the one with the greatest resistance will have the greatest potential difference across it, and vice versa.

Parallel Circuits

Components connected in parallel are connected separately in their own loop going from one terminal of the battery to the other.

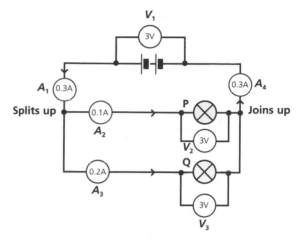

- The amount of current that passes through each component depends on the resistance of each component. The greater the resistance, the smaller the current. In the circuit above, bulb Q has half the resistance of bulb P, so twice as much current flows through it.
- The total current from the battery is equal to the sum of the current through each of the parallel components, i.e. $A_1 = A_2 + A_3 = A_4$. In the circuit above, 0.3A = 0.1A + 0.2A = 0.3A.
- The current is smallest through the component with the greatest resistance.

> **HT** The same voltage causes more current to flow through a smaller resistance than a bigger one.
>
> The potential difference across each component is equal to the potential difference of the battery.
>
> The current through each component is the same as if it was the only component present. For example, a circuit with a battery and a bulb has a 1 amp current. If a second identical component is added in parallel, the current through each would be 1 amp. The total current through the battery will increase.

Electromagnetic Induction

In electromagnetic induction, **movement produces voltage**. If a wire or a coil of wire cuts through the lines of force of a magnetic field (or vice versa), then a voltage is induced (produced) between the ends of the wire. If the wire is part of a complete circuit, a current will be induced.

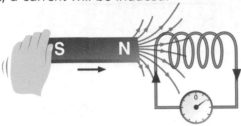

Moving the magnet into the coil induces a current in one direction. A current can be induced in the opposite direction by:
- moving the magnet out of the coil

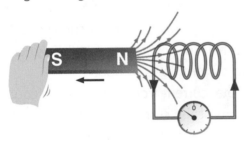

- moving the other pole of the magnet into the coil.

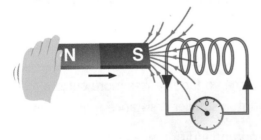

Both of these involve a magnetic field being cut by a coil of wire, creating an induced voltage. If there is no movement of the magnet or coil there is no induced current, because the magnetic field is not being cut.

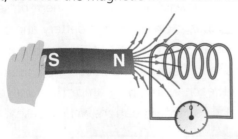

The Electric Generator

Mains electricity is produced by generators. Generators use the principle of electromagnetic induction to generate electricity by rotating a magnet inside a coil.

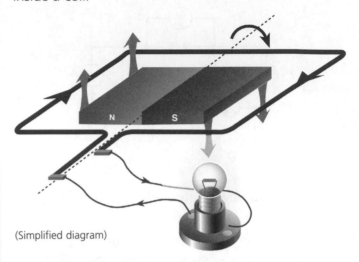

(Simplified diagram)

The size of the induced voltage follows the same rules as with the magnet and coil. The voltage will be increased by:
- increasing the speed of rotation of the magnet
- increasing the strength of the magnetic field
- increasing the number of turns on the coil
- placing an iron core inside the coil.

HT As the magnet rotates, the **voltage** induced in the coil changes direction and size, as seen below.

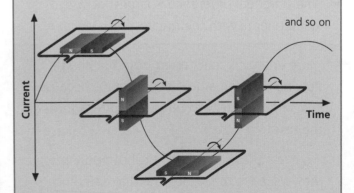

The induced current reverses its direction of flow every half turn of the magnet and is, therefore, alternating current.

Power

When charge flows through a component, energy is transferred to the component. Power, measured in watts (W), is a measure of how much energy is transferred every second, i.e. **the rate of energy transfer**.

Electrical power can be calculated using the following equation:

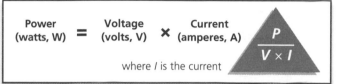

Power (watts, W)	=	Voltage (volts, V)	×	Current (amperes, A)

$$\frac{P}{V \times I}$$

where *I* is the current

Example

An electric motor works at a current of 3A and a potential difference of 24V. What is the power of the motor?

Power = Potential difference × Current

= 24V × 3A

= **72W**

HT The power formula can be rearranged using the formula triangle to work out the potential difference or current.

Example

A 4W light bulb works at a current of 2A. What is the potential difference?

$$\text{Potential difference} = \frac{\text{Power}}{\text{Current}}$$

$$= \frac{4W}{2A}$$

$$= \textbf{2V}$$

Transformers

Iron core

Primary coil   Secondary coil

Transformers are used to change the voltage of an alternating current. They consist of two coils, called the **primary** and **secondary** coils, wrapped around a soft iron core.

HT When two coils of wire are close to each other, a changing magnetic field in one coil can induce a voltage in the other. Alternating current flowing through the primary coil creates an alternating magnetic field. This changing field then induces an alternating voltage across the secondary coil, which causes an alternating current through the secondary coil.

The amount by which a transformer changes the voltage depends on the number of turns on the primary and secondary coils. If the number of turns on each coil is the same, the voltage does not change. If there are more turns on the secondary coil, the voltage goes up; if there are fewer turns, the voltage goes down.

You need to be able to use this equation:

Voltage on primary coil, V_p		Number of turns on primary coil, N_p
Voltage on secondary coil, V_s	=	Number of turns on secondary coil, N_s

Example

A transformer has 1000 turns on the primary coil and 200 turns on the secondary coil.

If a voltage of 250V is applied to the primary coil, what is the voltage across the secondary coil?

Using our formula: $\dfrac{V_p}{V_s} = \dfrac{N_p}{N_s}$

$$\frac{250}{V_s} = \frac{1000}{200}$$

$$250 = 5V_s$$

$$V_s = \frac{250}{5}$$

$$V_s = \textbf{50V}$$

How an a.c. Generator Works

In simple terms, an a.c. generator works by placing a coil of wire within a magnetic field and **rotating** it.

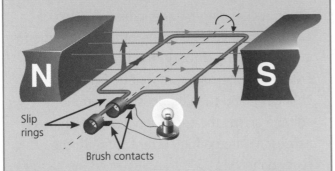

Slip rings

Brush contacts

As it rotates and cuts through the magnetic field, a voltage is induced across the coil and a current is induced in the coil. One side of the coil moves up during one half turn, and then down during the next half. This means that the current reverses direction, or **alternates**, every half turn (after it has turned through 180°).

Each end of the coil is attached to a separate metal ring, called a **slip ring**, which rotates with it. The generator can be connected to an external circuit (e.g. to power a light) using brush contacts. These brushes are spring-loaded so that they continuously push against the slip rings and the circuit remains complete. They need to be replaced regularly because they gradually wear away.

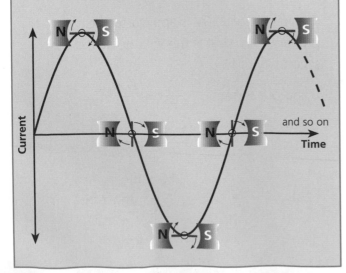

Types of Current

A **direct current** (d.c.) always flows in the same direction. Cells and batteries supply direct current.

An **alternating current** (a.c.) changes the direction of flow back and forth continuously. The number of complete cycles per second is called the **frequency**, and for UK mains electricity this is 50 cycles per second (hertz, Hz).

In the UK, the mains supply has a voltage of about 230 volts which, if it is not used safely, can kill.

We use a.c. instead of d.c. for our electrical supply for different reasons:

- a.c. is easier to generate and distribute over large distances within the National Grid. When d.c. is transmitted over large distances, it loses a lot of electrical energy within the wires due to heat, etc. This would mean that the further you live from the power station, the less energy you would get.

- a.c. can be increased and decreased using transformers that reduce energy loss due to heat. Therefore a.c. can be transmitted with hardly any power loss. d.c. cannot be used with transformers. At the power station, transformers step up the voltage to between 150 000V and 400 000V. Before electricity is consumed by the domestic user, transformers step down the voltage of the electricity to a level that is safe to use, e.g. 230V.

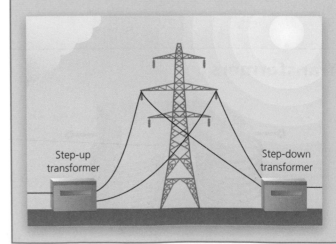

Step-up transformer

Step-down transformer

The Principles of the Motor Effect

In the motor effect, **current produces movement**. When a conductor (wire) carrying an electric current is placed in a magnetic field, the magnetic field formed around the wire interacts with the permanent magnetic field causing the wire to experience a force, which makes it move. This force acts at right angles to both the current and magnetic field. (The force is shown by green arrows in the diagrams below.)

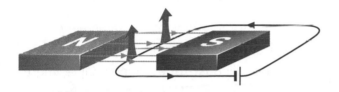

The **size** of the force on the wire can be increased by:

* increasing the size of the current (e.g. having more cells)

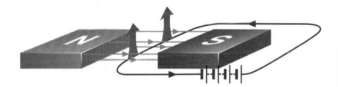

* increasing the strength of the magnetic field (e.g. having stronger magnets).

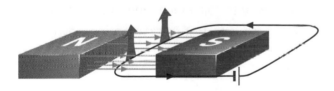

The **direction** of the force on the wire can be reversed by:

* reversing the direction of flow of the current (e.g. turning the cell around)

* reversing the direction of the magnetic field (e.g. swapping the magnets around).

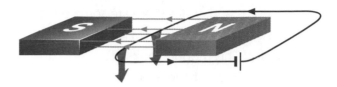

The wire will not experience a force if it is parallel to the magnetic field.

The Direct Current Motor

Electric motors rely on the principle of the motor effect. They form the basis of a vast range of electrical devices both inside and outside the home.

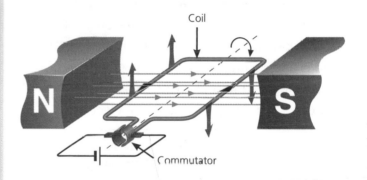

As a current flows through the coil, a magnetic field is formed around the coil, creating an electromagnet. This magnetic field interacts with the permanent magnetic field that exists between the two poles, North and South. A force acts on both sides of the coil, which rotates the coil to give us a very simple motor.

To make the motion continuous, a commutator is used. The commutator makes the direction of the current reverse every half turn, so the motor keeps turning in the same direction.

A motor can therefore make something move. Motors are used to move the drum in a washing machine, spin the disc in a DVD player and a hard disk drive, and turn the wheels of an electric car. Tiny electric motors drive the wheels in toy cars, and huge motors drive electric trains.

Module P6 (Radioactive Materials)

Some materials are radioactive because of the structure of the atoms they contain. Radiation is all around us and, while there are some dangers and hazards to address, certain radiations are very useful. This module looks at:

- the structure of the atom
- nuclear fusion
- alpha, beta and gamma radiation
- half-life
- background radiation
- uses of radiation
- nuclear fission and nuclear power stations.

Atoms and Elements

All elements are made of **atoms**; each element contains only one type of atom. All atoms contain a nucleus and electrons. The nucleus is made from protons and neutrons with the one exception of hydrogen (the lightest element), which has no neutrons – just one proton and one electron.

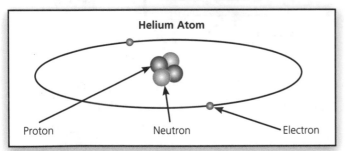

Helium Atom

Proton   Neutron   Electron

Some elements give out **ionising radiation** all the time; we call these elements **radioactive**. Neither chemical reactions nor physical processes (e.g. heating) can change the radioactive behaviour of a radioactive substance.

 An atom has a nucleus made of protons and neutrons. Every atom of a particular **element** always has the same number of protons (if it contained a different number of protons it would be a different element). For example…

- hydrogen atoms have one proton
- helium atoms have two protons
- oxygen atoms have eight protons.

However, some atoms of the same element can have different numbers of neutrons – these are called **isotopes**. For example, these are three isotopes of oxygen:

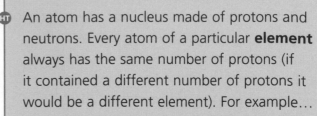

| Oxygen-16 | Oxygen-17 | Oxygen-18 |
| has eight neutrons | has nine neutrons | has ten neutrons |

All three of these isotopes have eight protons.

Ionising Radiation

There are three types of radiation that can be given out by radioactive materials. These radiations are **alpha**, **beta** and **gamma**.

Different radioactive materials will give out any one, or a combination, of these radiations.

An easy way to tell which type of radiation you are dealing with is to test its penetrating power.

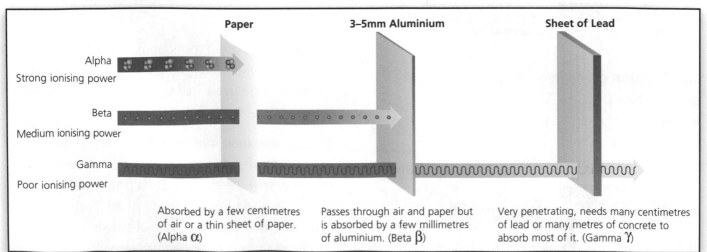

| Paper | 3–5mm Aluminium | Sheet of Lead |

Alpha — Strong ionising power

Beta — Medium ionising power

Gamma — Poor ionising power

Absorbed by a few centimetres of air or a thin sheet of paper. (Alpha α)

Passes through air and paper but is absorbed by a few millimetres of aluminium. (Beta β)

Very penetrating, needs many centimetres of lead or many metres of concrete to absorb most of it. (Gamma γ)

Nuclear Fusion

At the beginning of the 20th century, discoveries about the nature of the atom and nuclear processes began to help answer the mystery of where the Sun's energy comes from.

In 1911, there was a ground-breaking experiment – the Rutherford–Geiger–Marsden alpha particle scattering experiment. In this experiment, a thin gold foil was bombarded with alpha particles. The effect on the alpha particles was recorded, and these observations provided the evidence for our current understanding of atoms.

Most alpha particles were seen to **pass straight through** the gold foil. This would indicate that gold atoms were composed of large amounts of open space. However, some particles were **deflected** slightly and a few were even **deflected back** towards the source. This would indicate that the alpha particles passed close to something positively charged within the atom and were repelled by it.

These observations brought Rutherford and Marsden to conclude that:

- gold atoms, and therefore all atoms, consist of largely empty space with a small, positive region called the **nucleus**
- the nucleus is positively charged
- the electrons are arranged around the nucleus with a great deal of space between them.

HT Rutherford's data allowed him to identify a small positive nucleus, but the limited data did not explain its structure. Later experiments revealed the presence of positive protons and neutral neutrons in the nucleus held together by the short-range **strong nuclear force**. Protons normally repel each other (because they have the same charge), but the strong nuclear force balances the repulsive electrostatic force between the protons.

The strong nuclear force between protons and neutrons can also cause hydrogen nuclei (or protons) that are close enough to each other to fuse into helium nuclei. This process releases large amounts of

energy and is known as **nuclear fusion**. It is the same process that occurs inside the Sun.

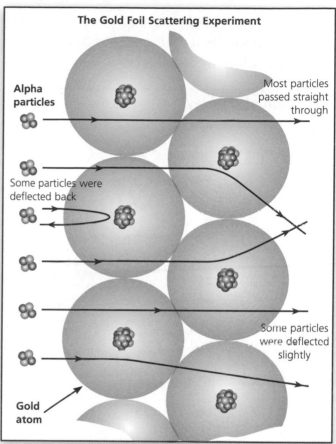

The Gold Foil Scattering Experiment

Alpha particles

Most particles passed straight through

Some particles were deflected back

Some particles were deflected slightly

Gold atom

HT ## Einstein's Equation

Perhaps one of Albert Einstein's greatest insights was to realise that matter and energy are really different forms of the same thing. He quantified this brilliantly with the famous equation:

$$E = mc^2$$

E = The energy produced
m = The mass lost
c = The speed of light (in a vacuum) 300 000 000 m/s

This same equation can be used to calculate the energy released during nuclear fusion and fission.

If two hydrogen nuclei are forced together, one of the protons will change into a neutron, resulting in the formation of a deuterium nucleus. The deuterium nucleus has a lower mass than the proton and neutron separately. This 'mass defect' is converted into energy and can be calculated using $E = mc^2$.

Radioactive Decay

The emission of ionising radiation occurs because the nucleus of an unstable atom is decaying. The type of decay depends on why the nucleus is unstable; the process of decay helps make the atom become more stable. During decay, the number of protons in the atom may change. If this happens, the element changes from one type to another.

Alpha (α) decay

Unstable parent ➡ New daughter ➕ Alpha particle

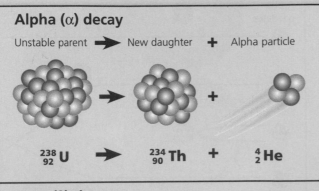

$$^{238}_{92}U \rightarrow \ ^{234}_{90}Th \ + \ ^{4}_{2}He$$

The original atom decays by ejecting an alpha (α) particle from the nucleus. This particle is a helium nucleus: a particle made up of two protons and two neutrons. With alpha decay a new atom is formed. This new atom has two protons and two neutrons fewer than the original.

Beta (β) decay

Unstable parent ➡ New daughter ➕ Beta particle

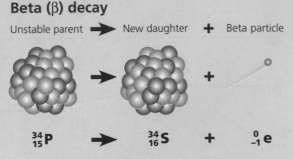

$$^{34}_{15}P \rightarrow \ ^{34}_{16}S \ + \ ^{0}_{-1}e$$

The original atom decays by changing a neutron into a proton and an electron. This high-energy electron, which is now ejected from the nucleus, is a beta (β) particle. With beta decay a new atom is formed. This new atom has one more proton and one less neutron than the original.

Gamma (γ) decay

Since a gamma ray has no mass, there is no change to the 'parent' element.

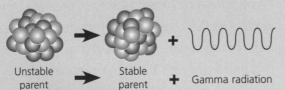

Unstable parent ➡ Stable parent ➕ Gamma radiation

After α or β decay, a nucleus sometimes contains surplus energy. It emits this as gamma radiation, which is very high frequency electromagnetic radiation. During gamma decay, only energy is emitted. This does not change the type of atom.

Background Radiation

Radioactive elements are found naturally in the environment. The radiation produced by these sources contributes to the overall background radiation.

There is nothing we can do to prevent ourselves from being **irradiated** and **contaminated** by background radiation, but the level of background radiation in most places is so low that it is nothing to worry about. There is, however, a correlation between certain cancers and living for many years in areas where the underlying rock is granite.

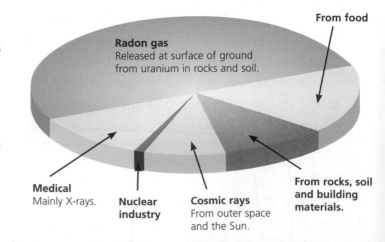

Radon gas
Released at surface of ground from uranium in rocks and soil.

From food

Medical
Mainly X-rays.

Nuclear industry

Cosmic rays
From outer space and the Sun.

From rocks, soil and building materials.

Half-life

The **activity** of a substance is a measure of the amount of radiation given out per second.

When a radioactive atom decays it becomes less radioactive and its activity drops. The **half-life** of a substance is the time it takes for the radiation emitted by a radioactive material to halve. The half-life of a radioactive material can range from a few seconds to millions of years.

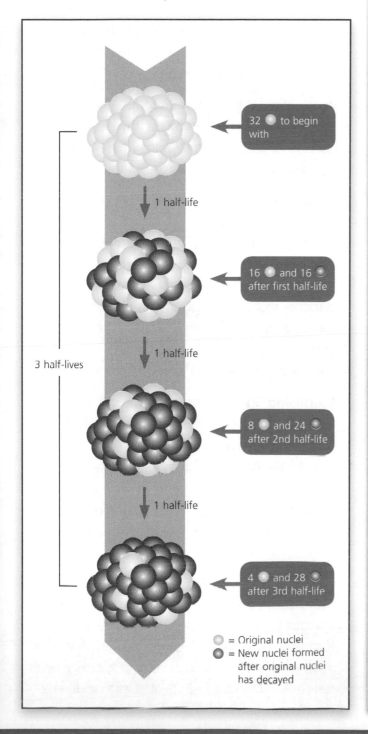

3 half-lives

32 ● to begin with

16 ● and 16 ● after first half-life

1 half-life

8 ● and 24 ● after 2nd half-life

1 half-life

4 ● and 28 ● after 3rd half-life

○ = Original nuclei
● = New nuclei formed after original nuclei has decayed

Half-life and Safety

All radioactive substances become less radioactive as time passes. A substance would be considered safe once its activity had dropped to the same level emitted as background radiation, which is a dose of around 2 millisieverts per year or 25 counts per minute with a standard detector.

Some substances decay quickly and could be safe in a very short time. Substances with a long half-life remain harmful for millions of years.

ᴴᵀ Half-life Calculations

The half-life can be used to calculate how old a radioactive substance is, or how long it will take to become safe.

Example

If a sample has an activity of 800 counts per minute and a half-life of 2 hours, how many hours will it take for the activity to reach the background rate of 25 counts per minute?

We need to work out how many half-lives it takes for the sample of 800 counts to reach 25 counts.

1. $\frac{800}{2} = 400$

2. $\frac{400}{2} = 200$

3. $\frac{200}{2} = 100$

4. $\frac{100}{2} = 50$

5. $\frac{50}{2} = 25$

It takes 5 half-lives to reach a count of 25, and each half-life takes 2 hours, so it takes:

5 × 2 hours = **10 hours**

Ionisation

When radiation interacts with neutral atoms or molecules, the atoms or molecules may become charged due to electrons being knocked out of their structure. This alters their structure, leaving them as charged particles called **ions**. Alpha, beta and gamma radiation are therefore known as ionising radiation and can damage molecules in healthy cells, which results in the death of the cell.

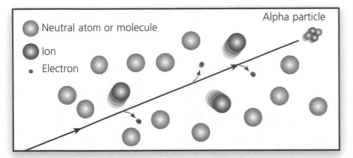

Cell Damage

When living cells absorb radiation, damage can occur in different ways:

- Ionising radiation can damage cells, causing ageing of the skin.
- Ionising radiation can cause mutations in the nucleus of a cell, which can lead to cancer.
- Different amounts of exposure can cause different effects, e.g. high-intensity ionising radiation can kill cells, leading to radiation poisoning.

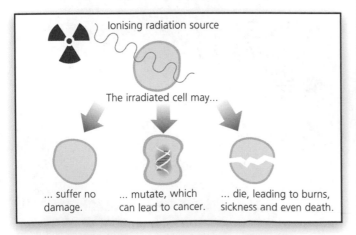

HT Ionising radiation is able to break molecules into bits (called ions), which can then take part in other chemical reactions.

Uses of Radiation

Although using ionising radiation can be dangerous, there are many beneficial uses:

- to treat cancer
- to sterilise surgical instruments and food
- as a tracer in the body for medical procedures.

HT Cancer Treatment

High-energy gamma rays can be used to kill cancer cells. However, ionising radiation can damage living cells too, so the radiation has to be carefully targeted from different angles to minimise the damage to healthy cells.

Radioactive iodine can be used to target thyroid cancer. Iodine is needed by the thyroid, so it collects naturally in the thyroid, where it gives out beta radiation and kills the cancer cells.

In both of these examples there is a danger of damage to healthy cells, so the doctors need to carefully weigh the risks against the benefits. When deciding on the dose of radiation to use, the **ALARA** (as low as reasonably achievable) principle should be applied. This means that measures should be taken to make the risks as small as possible, while still providing the benefits and taking into account all the implications.

Sterilising Surgical Instruments

Surgical instruments must not have any bacteria on them. Bacteria are living cells, so are susceptible to damage from ionising radiation. Exposing surgical instruments in sealed bags to gamma radiation results in the death of the bacterial cells, and therefore sterilisation of the surgical instruments. The instruments remain sterile until they are ready to be used.

Sterilising Food

Irradiating food kills any bacteria that would cause the food to go off. Radiation treatment is only allowed on a few foods in the UK and these have to carry a label stating that they have been treated with radiation.

ⓗⓣ Tracers in the Body

Doctors use radioactive chemicals called tracers to help detect damage to the internal organs. Once the tracer enters the body, it builds up in the damaged or diseased part of the body. Radiation detectors can then be used to detect where the problem is. These can be linked to computers, which produce an image showing the distribution of the radioactive chemical. Problem areas are highlighted by a high concentration.

Because it is being used inside the body, the radioactive tracer must be non-toxic. It also needs to have a short half-life, so that it breaks down quickly after use. Gamma and beta sources are used because they pass out of the body easily. An alpha source is never used because it would be quickly absorbed and cause damage.

Technetium-99 is a gamma emitter often used for this purpose. It has a half-life of 6 hours. This gives doctors enough time to detect the problem, but ensures that the radiation decreases to a safe level quickly.

Dangers of Radiation

New scientific advances often create an element of risk. The transportation and application of radioactive substances is carefully controlled by government rules and regulations to minimise the risk to the general public. However, people working in the nuclear industry, medical physics (X-rays, etc.) and many other areas often have to work with radioactive materials. They can become irradiated or contaminated by the materials, which could lead to serious health problems or even death. Therefore, the exposure that these people are subjected to needs to be carefully monitored.

The different types of radiation carry different risks:

- Alpha is the most dangerous if the source is inside the body; all the radiation will be absorbed by cells in the body.
- Beta is the most dangerous if the source is outside the body because, unlike alpha, it can penetrate the outer layer of skin and damage the internal organs.
- Gamma can cause harm if it is absorbed by the cells in the body, but it is weakly ionising and can pass straight through the body, causing little damage.

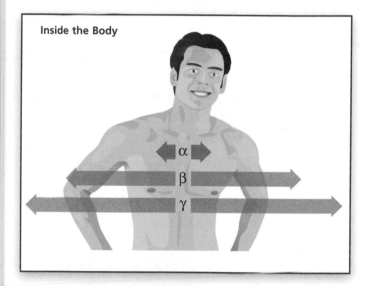

Inside the Body

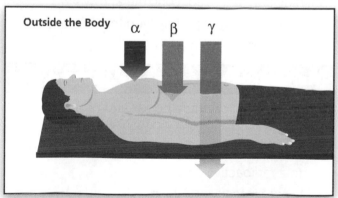

Outside the Body

The sievert is a measure of a radiation dose's potential to do harm to a person. It is based on both the type and amount of radiation absorbed.

A 5 sievert dose is a 5 sievert dose regardless of the type of radiation absorbed. This makes the sievert a very useful unit for radiation safety measures.

Radiation Protection

The transportation and application of radioactive substances is carefully controlled by government rules and regulations to minimise the risks to the general public. Authorised personnel who handle radioactive substances or operate machinery follow strict guidelines. These vary depending on the risk factor involved. The table below lists some measures that can be taken to reduce/prevent exposure to different types of radiation.

External Exposure – Beta and Gamma

- Minimise time of exposure.
- Maximise distance from the source of radiation.
- Wear protective clothing (and remove it before leaving a restricted area).
- Avoid direct handling of radioactive materials, i.e. use implements like forceps and tongs.
- Use protective shields, screens and containers.
- Use instruments that can detect levels of radiation, e.g. Geiger-Müller counter or liquid scintillation counter.
- Use materials that can provide a shield against radiation, e.g. lead, concrete and water.
- Wear a film badge that monitors the degree of exposure to radiation.

Internal Exposure – Alpha, Beta and Gamma

- Wear chemical fume hoods and protective masks to prevent inhalation.
- Never consume or store food and drink close to a radioactive source.
- Wear protective clothing.
- Ensure that any cuts or wounds are sealed up.
- Minimise amount of radioactive material to be handled.

People who work in the nuclear industry, such as radiographers and nuclear power plant technicians, are regularly exposed to radiation. They often wear a badge to monitor their degree of exposure. The badge contains photographic film, which (after developing) becomes darker the more it is exposed to radiation. In this way, radiation exposure can be carefully monitored.

Other physical barriers are used to protect people from ionising radiation:

- Radiographers use lead screens to prevent exposure.
- Nuclear reactors are encased in thick lead and concrete to prevent radiation escaping into the environment.
- Nuclear technicians going into areas of high levels of radiation must wear a radiation suit made from materials that act as a shield against the radiation.

Nuclear Fission

In a chemical reaction it is the electrons that bring about the change. The elements involved remain the same but join up in different ways.

A fission reaction takes place in the nucleus of the atom and different elements are formed. A neutron is absorbed by a large and unstable **uranium** or **plutonium** nucleus. This splits the nucleus into two, roughly equal-sized, smaller nuclei and releases energy and more neutrons.

A fission reaction releases far more energy than that released from a chemical reaction involving a similar mass of material. Once fission has taken place the neutrons released can be absorbed by other nuclei and further fission reactions can take place. This is called a **chain reaction.**

A chain reaction occurs when there is enough fissile material to prevent too many neutrons escaping without being absorbed. This is called critical mass and ensures every reaction triggers at least one further reaction (see opposite).

The Nuclear Reactor

Nuclear power stations use fission reactions to generate the heat needed to boil water into steam. The reactor controls the **chain reaction** so that the energy is released at a steady rate.

Fission occurs in the **fuel rods** and causes them to become very hot.

The **coolant** is a liquid that is pumped through the reactor. The coolant heats up and is then used in the heat exchanger to turn water into steam.

Control rods, made of boron, absorb neutrons, preventing the chain reaction getting out of control. Moving the control rods in and out of the reactor core changes the amount of fission that takes place.

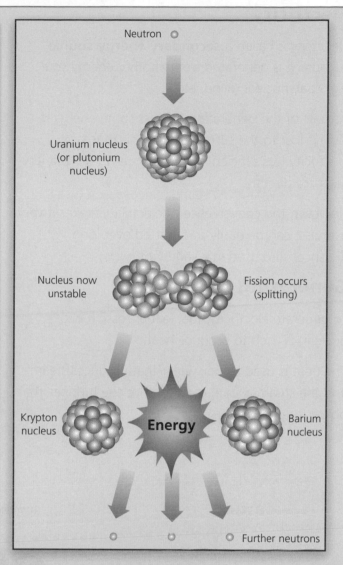

Neutron

Uranium nucleus
(or plutonium
nucleus)

Nucleus now
unstable

Fission occurs
(splitting)

Krypton
nucleus

Energy

Barium
nucleus

Further neutrons

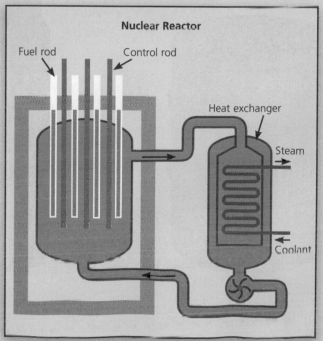

Nuclear Reactor

Fuel rod   Control rod

Heat exchanger

Steam

Coolant

Electricity

Electricity is called a **secondary energy source** because it is generated from another energy source e.g. coal, nuclear, wind, etc.

As part of the generation process some energy is always lost to the surroundings. This makes electricity less efficient than when using the primary resource directly.

However, the convenience of electricity makes it very useful. It can be easily transmitted over long distances and used in a variety of ways.

Generating Electricity

To generate electricity, fuel (either fossil fuel or nuclear) is used to produce heat.

The heat is used to boil water that produces steam, and the steam is then used to drive the turbines that power the generators.

The electricity produced in the generators is sent to a transformer and then to the National Grid, from where we can access it in our homes.

In a **nuclear** power station, the energy is released due to changes in the nucleus of radioactive substances. Nuclear power stations do not produce carbon dioxide but they do produce **radioactive waste**. This nuclear waste is categorised into three types:

- **High-level waste** (HLW) – very radioactive waste that has to be stored carefully. Fortunately, only small amounts are produced and its activity decreases quickly, so it is put into short-term storage.
- **Intermediate-level waste** (ILW) – not as radioactive as HLW but it does remain radioactive for thousands of years. The amount produced is increasing; deciding how to store it safely and permanently is a problem. At the moment most ILW is mixed with concrete and stored in big containers, but this is not a permanent solution.
- **Low-level waste** (LLW) – only slightly radioactive waste that is sealed and placed in landfills.

Low-level Waste

Sealed in steel drums

Intermediate-level Waste

Sealed in concrete, surrounded by steel

High-level Waste

Sealed in glass, surrounded by steel

1 Which of the following statements about ionising radiation are true? Put ticks (✓) in the boxes next to the two correct answers.

    A Ionising radiation can break molecules into bits called ions. ☐

    B X-rays are ionising radiation; gamma rays are not. ☐

    C Ionising radiation cannot break molecules into bits called ions. ☐

    D X-rays and gamma rays are ionising radiations. ☐ **[2]**

2 (a) A car has a mass of 800kg and is travelling at 20m/s. Calculate the momentum of the car. Include units in your answer. **[3]**

    (b) How will the momentum change if the car is loaded with bricks but still travels at 20m/s? **[1]**

    (c) How will the momentum change if the car speeds up? **[1]**

3 (a) The following circuit has two identical bulbs. Write down the missing values for:

    (i) A_1 **[1]**

    (ii) A_3 **[1]**

    (iii) V_2 **[1]**

    (iv) V_3 **[1]**

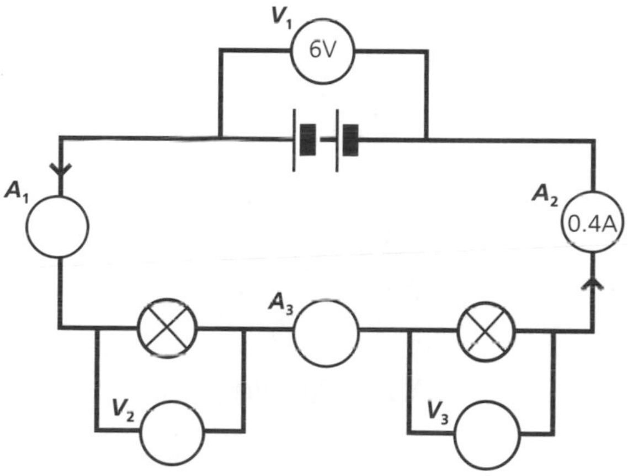

    (b) Draw straight lines between the boxes to join the circuit symbols with their correct meaning. **[3]**

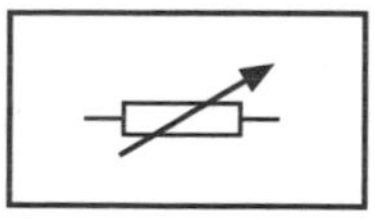

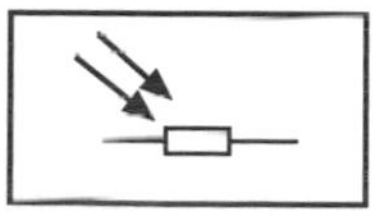

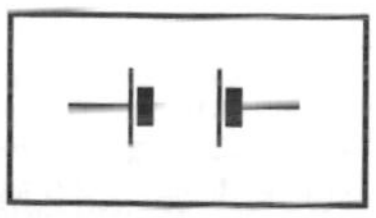

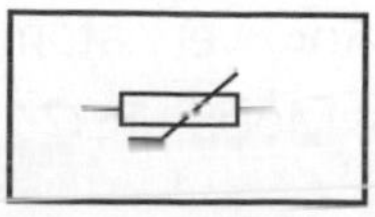

	Battery
	Thermistor
	Variable resistor
	LDR

Exam Practice Questions

④ Name three sources that contribute to background radiation. **[3]**

⑤ What is the name of the process that releases energy in a nuclear power station? **[1]**

⑥ A school bus sets off from school and drives to its first stop. Describe the motion of the bus, using the distance–time graph below. **[3]**

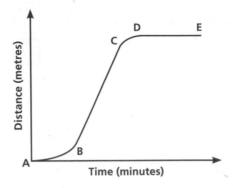

⑦ This question is about nuclear power.

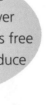

I'm the manager of the Super Fuels Nuclear Power Company. Nuclear energy is free energy and it doesn't produce any pollution.

I live opposite the power station, and I think nuclear power is dangerous, like a ticking time bomb.

I work for the electricity board. I think that nuclear power is the only alternative to fossil fuels.

Mr Benbow   **Miss Manning**   **Mr Holmes**

(a) For two of the above people, explain why their point of view may be unreliable. **[1]**

(b) For one of the above people, explain why their point of view may be wrong. **[1]**

⑧ In 1911, Rutherford and Marsden carried out an experiment that disproved the 'plum pudding' theory that described the structure of the atom. They fired alpha particles at gold foil in a vacuum and recorded the paths of the alpha particles.

Describe the observations made and the conclusions drawn from the experiment. **[6]**

✏ *The quality of written communication will be assessed in your answer to this question.*

HT

⑨ Potassium-40 decays by beta emission to argon-40. A sample of Moon rock that originally contained no argon-40 atoms now contains 15 atoms of potassium-40 for every atom of argon-40. The half-life of potassium-40 is 1 billion years. How old is the Moon rock? **[2]**

Module P7 (Further Physics – Studying the Universe)

The study of the Universe has resulted in a greater understanding of our planet. This module looks at:

- what the objects in the night sky are
- how far away objects in space are
- the life cycle of stars
- telescopes and how they work.

Looking Into Space

The Earth fully rotates on its axis once in just under 24 hours. We cannot feel the Earth spinning, but it is due to this rotation that the stars **appear** to move **east–west** across the sky once in just under 24 hours.

The Sun and Moon also appear to travel east–west across the sky. Their motion and the time they take to cross the sky are affected by their orbits. In the case of the Sun, it appears to travel across the sky once every 24 hours.

HT The planets also appear to travel east–west across the sky. Their motion and the time they take to cross the sky are affected by their orbits.

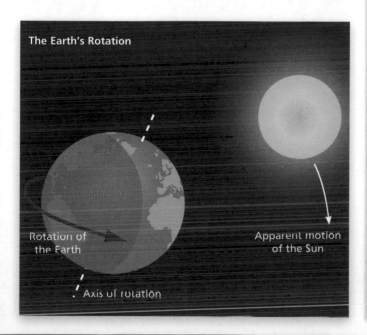

The Earth's Rotation

Rotation of the Earth

Apparent motion of the Sun

Axis of rotation

HT The Earth and the Sun

A **sidereal day** is the time it takes for the Earth to rotate 360° on its axis.

A **solar day** is the time from noon on one day, to noon on the next day, i.e. 24 hours.

As the Earth rotates once on its axis, it is also **orbiting** the Sun. It is this **orbiting** movement that results in a sidereal day being shorter than a solar day.

A sidereal day is 23 hours and 56 minutes – four minutes shorter than a solar day.

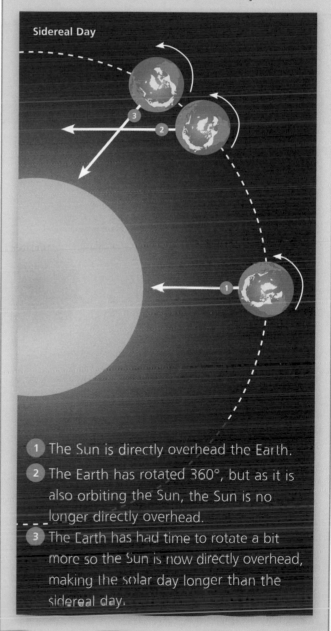

Sidereal Day

1. The Sun is directly overhead the Earth.
2. The Earth has rotated 360°, but as it is also orbiting the Sun, the Sun is no longer directly overhead.
3. The Earth has had time to rotate a bit more so the Sun is now directly overhead, making the solar day longer than the sidereal day.

The Position of the Stars

Due to the orbiting movement of the Earth around the Sun, an observer looking at the night sky from the Earth can see different stars at different times of the year, depending on the Earth's position in relation to the Sun's position.

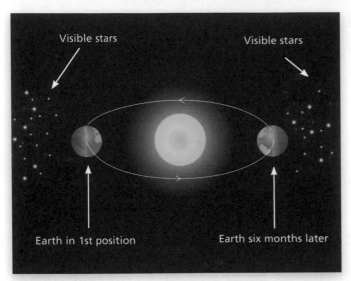

The Planets

Mercury, Venus, Mars, Saturn and Jupiter are all planets that can be seen from Earth with the naked eye. These planets look similar to stars, but they change their positions in complicated patterns when compared with the background of fixed stars.

It is these complicated patterns that provided some of the first evidence that the planets, including the Earth, orbit the Sun.

ⓗ Retrograde Motion

We can use the observations of Mars as an example to show how the planets change their position against the background stars.

Earth is closer to the Sun than Mars, so Earth's orbit of the Sun takes less time than Mars's orbit of the Sun. If Mars is observed over a long enough period of time, it can be seen to move compared with the background stars.

However, approximately once every two years, Mars appears to go back on itself, e.g. it goes east to west rather than west to east. This is called **retrograde motion**.

Plotting Astronomical Objects

When astronomers look into space, they describe the position of objects in terms of the angles of **declination** and **right ascension**. These angles describe the positions of the stars relative to a fixed point on the equator.

ⓗ The Celestial Sphere

The **celestial sphere** is an imaginary sphere enclosing the Earth that allows astronomers anywhere in the world to find a particular star or constellation. The celestial sphere can be used to find stars if you know the star's declination and right ascension. Right ascension is measured in hours; the celestial sphere is split up into 24 hours of right ascension as there are 24 hours in a day.

A star with a **positive declination** will be visible from the **northern hemisphere**. A star with a **negative declination** will be visible from the **southern hemisphere**.

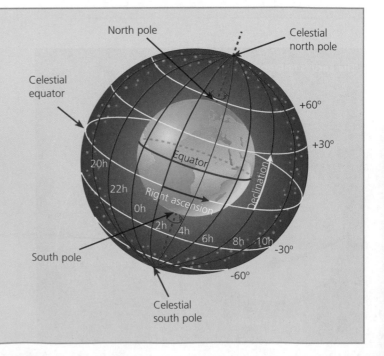

Retrograde Motion (Cont.)

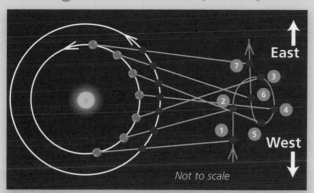

Not to scale

As the Earth is closer to the Sun than Mars and is travelling faster, it can 'catch up' and undertake Mars as it orbits the Sun. As the Earth goes past Mars, Mars appears to go back on itself compared with the stars in the night sky. Mars appears to 'wander' across the sky and was one of the first pieces of evidence that the planets orbit the Sun and not the Earth, as previously thought. The word 'planet' comes from the ancient Greek 'wanderer'.

The Earth and the Moon

While the Earth is rotating on its axis, the Moon is orbiting the Earth in the same direction. Due to this orbiting movement, the Moon appears to travel east–west across the sky in a little over 24 hours.

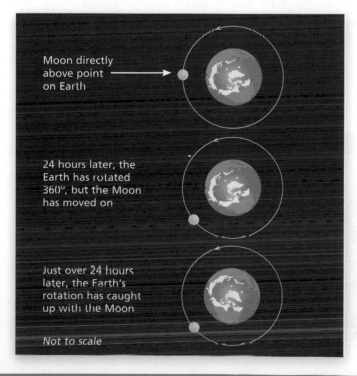

Not to scale

The Lunar Cycle

The **lunar cycle** describes the Moon's appearance during its 28-day orbit of the Earth.

The Moon's shape has nothing to do with the shadow of the Earth, but is due to the part of the Moon that is **visible** from the Earth.

The Moon is visible from Earth because we can see the light from the Sun reflected from it.

This means that the side of the Moon facing away from the Sun appears dark.

During the Moon's orbit around the Earth, we can see different faces of the Moon:

- dark face (new Moon)
- light face (full Moon)
- all the points in between the new Moon and full Moon.

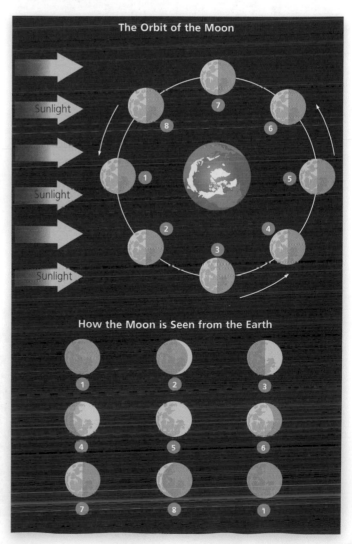

Eclipses

A **lunar eclipse** occurs when the Earth is between the Sun and the Moon. This results in the Earth casting a shadow on the Moon.

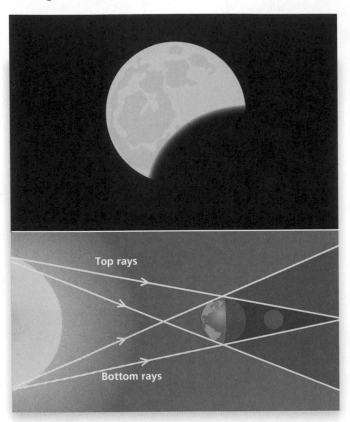

A **solar eclipse** occurs when the Moon passes between the Earth and the Sun. This can occur during a new Moon and results in the Moon casting a shadow on the Earth.

A **total solar eclipse** occurs when the Moon is directly in front of the Sun and completely obscures the Earth's view of the Sun.

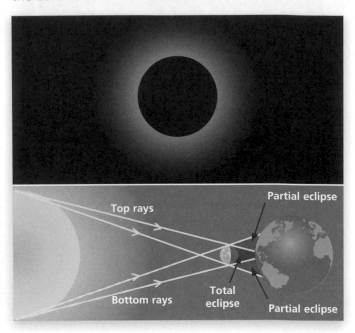

HT Eclipses do not occur every month because the Moon does not orbit the Earth in the same plane as the Earth orbits the Sun. The Moon's orbit is inclined 5° to that of the Earth's. Therefore, an eclipse can only occur when the Moon passes through the **ecliptic** (the apparent path the Sun traces out along the sky).

This is more likely to occur when the Moon is to the side of the Earth, rather than between the Earth and the Sun.

There are between two and five solar eclipses every year, but a total eclipse will only occur roughly every 18 months.

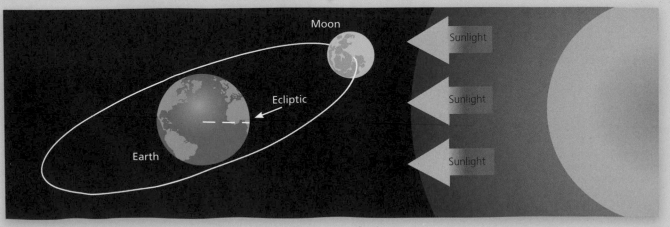

Behaviour of Waves

Waves, including light, water and sound waves, can be reflected, refracted or diffracted.

Reflection
Waves are reflected when a barrier is placed in their path. This effect can be seen in water waves.

Diffraction
When waves move through a narrow gap or past an obstacle, they spread out from the edges. This is diffraction. Diffraction is most obvious in two instances:

1. When the size of the gap is similar to the wavelength of the wave.
2. When the waves that pass obstacles have long wavelengths.

Light can be diffracted, but the waves need a small gap. The fact that light and sound can be diffracted provides evidence of their wave natures.

Reflection

Barrier

Incident wave

Reflected wave

Diffraction ❶

Slight diffraction ➡ Increased diffraction

Diffraction ❷

Slight diffraction ➡ Increased diffraction

Refraction of Light at an Interface

Light changes direction when it crosses an interface, i.e. a boundary between two transparent materials (media) of different densities, unless it meets the boundary at an angle of 90° (i.e. along the normal).

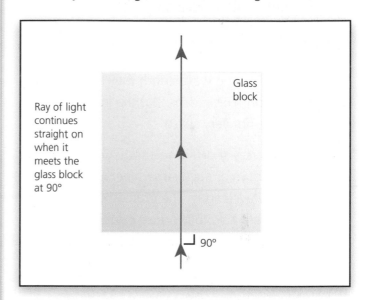

Glass block

Ray of light continues straight on when it meets the glass block at 90°

90°

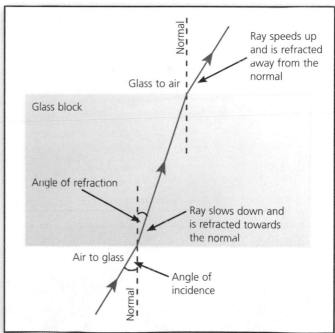

Normal

Ray speeds up and is refracted away from the normal

Glass to air

Glass block

Angle of refraction

Ray slows down and is refracted towards the normal

Air to glass

Angle of incidence

Normal

When the light passes from one medium to another, such as air to glass (more dense), its speed decreases, and this results in a change in direction by refraction. When the light leaves the glass and re-enters the air (less dense), it speeds up again.

Diffraction of Radiation

Radiation is diffracted by the aperture of a telescope. To produce a sharp image, the aperture must be much larger than the wavelength of the radiation.

Large radio telescopes that detect weak radio wave radiations can be built, but because radio waves have a long wavelength they are affected by diffraction. This means that the image produced is not very sharp.

Light has a very short wavelength. Optical telescopes have a much larger aperture than the light's wavelength. Therefore, the telescopes are able to produce a sharp image.

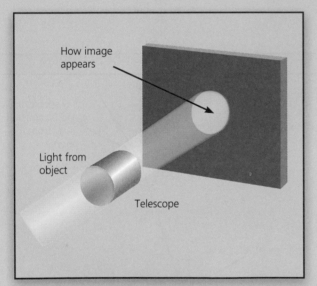

How image appears

Light from object

Telescope

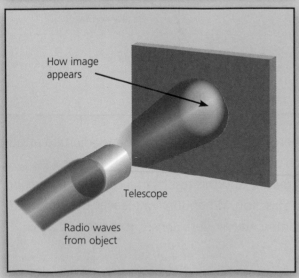

How image appears

Telescope

Radio waves from object

Convex Lenses

In a **convex** lens (also called a **converging lens**), rays of light are bent **inwards** as they pass through the lens. If the rays of light entering the lens are parallel, the rays will be brought to a focus at the **focal point** (F).

A lens with a more curved surface is more powerful than a lens with a less curved surface made of the same material.

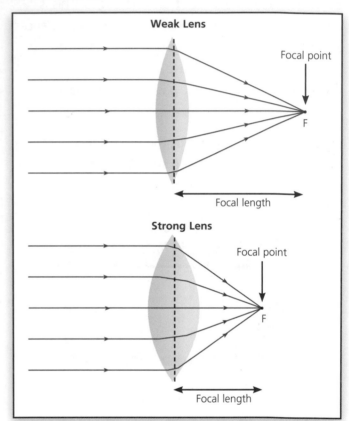

Weak Lens

Focal point

F

Focal length

Strong Lens

Focal point

F

Focal length

The power of a lens is measured in **dioptres** and can be calculated using the following formula:

$$\text{Power (dioptres)} = \frac{1}{\text{Focal length (metres)}}$$

Example

If a convex lens has a focal length of 10cm, calculate the power.

$$\text{Power} = \frac{1}{\text{Focal length}}$$

$$= \frac{1}{0.1\text{m}}$$

$$= \textbf{10 dioptres}$$

Ray Diagrams

You need to be able to interpret ray diagrams for converging lenses.

Example

Choose from the following options for each diagram:

- Magnified or diminished
- Real or virtual
- Upright or inverted.

Object Beyond 2F

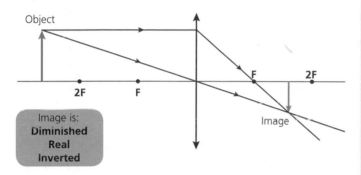

Image is:
**Diminished
Real
Inverted**

Object Between F and 2F

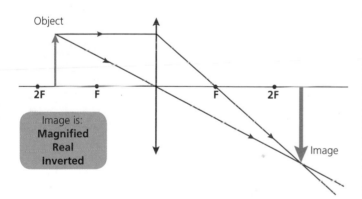

Image is:
**Magnified
Real
Inverted**

Object Between F and Lens

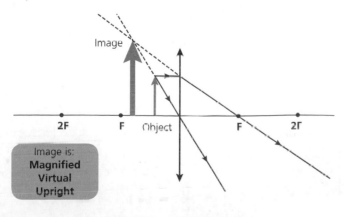

Image is:
**Magnified
Virtual
Upright**

HT You need to be able to draw ray diagrams for the formation of a **real image** from a distant point and a distant extended source.

Use this method to draw a ray diagram:

1. Draw a ray line (Ray 1) that runs from the top of the object parallel to the principal axis. At the middle of the lens, bend this ray inwards so it passes through the focal point.
2. Draw a second ray (Ray 2) that runs from the top of the object straight through the centre of the lens as it crosses the principal axis.
3. Draw a third ray (Ray 3) that runs from the top of the object through the focal point on the same side as the object. When the ray hits the centre of the lens, bend it to travel parallel to the principal axis.
4. The image is formed where these rays cross.

With a **distant extended source**, the images formed are real, **inverted** and **diminished** (smaller).

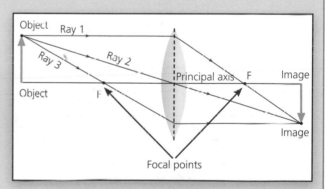

If the object crosses the principal axis, draw another ray going from the bottom of the object.

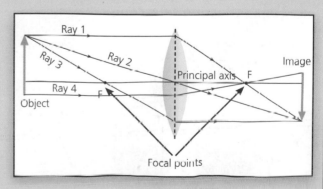

Telescopes

When looking into space, the objects are so far away that rays of light from them are effectively parallel. Therefore, we draw the rays of light entering telescopes as parallel rays.

A **simple refracting telescope** is made from two converging lenses of different powers. The **eyepiece lens** is a higher power lens than the **objective lens**.

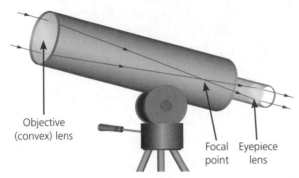

Objective (convex) lens

Focal point | Eyepiece lens

Astronomical telescopes will normally use **concave mirrors** for the objective lens instead of **convex lenses**. This allows them to be larger, so that they collect more light. A bigger lens and mirror, therefore, can see more distant objects. This diagram shows how a concave mirror focuses light rays.

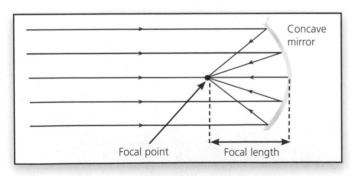

Concave mirror

Focal point | Focal length

The following diagram shows how a concave mirror is used in a **reflecting telescope**.

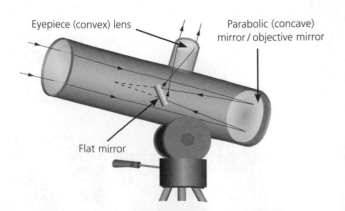

Eyepiece (convex) lens | Parabolic (concave) mirror / objective mirror

Flat mirror

If a distant object is magnified, the image appears closer than the object. Therefore, the angle made by ray lines entering the eye is greater. This increase in angle is called the **angular magnification** and makes the image appear bigger / closer.

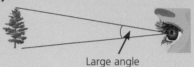

Near object

Large angle

Distant object

Small angle

Angular magnification (as seen through a telescope)

Apparent size of object

Large angle

The angular magnification of a telescope can be found if you know the focal length of the two lenses being used. You can use the following formula:

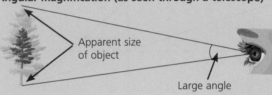

$$\text{Magnification} = \frac{\text{Focal length of objective lens}}{\text{Focal length of eyepiece lens}}$$

Example 1

The objective lens of a telescope has a focal length of 10m, and the eyepiece has a focal length of 2m. Calculate the magnification.

$$\text{Magnification} = \frac{\text{Focal length of objective lens}}{\text{Focal length of eyepiece lens}}$$

$$= \frac{10m}{2m}$$

$$= \times 5$$

Example 2

If the magnification of a telescope is ×10 and the objective lens has a focal length of 20cm, calculate the focal length of the eyepiece lens.

$$\frac{\text{Focal length of}}{\text{eyepiece lens}} = \frac{\text{Focal length of objective lens}}{\text{Magnification}}$$

$$= \frac{20}{10} = \textbf{2cm}$$

Measuring Distance Using Parallax

Parallax can be thought of as the apparent motion of an object against a background. However, it is actually the **motion of the observer** that causes the parallax **motion of an object**.

For example, in the diagram below, if an observer at position ➊ looks at the near star compared to the distant background, it appears to be at position B. If the observer looks at the same star six months later, (position ➋) the star appears to be at position A.

It looks as though the star has moved, but it is actually the movement of the Earth's orbit around the Sun that causes the observer to see this 'change in position'.

The distance to the stars is so great that we cannot observe parallax motion with the naked eye.

However, a simple way to observe parallax is if you hold your hand out in front of you with your thumb sticking up and alternately close one eye then the other.

Although your thumb appears to move, in reality you are just looking at it from a **different angle**.

The parallax angle (θ) of a star is half the angle moved against a background of distant stars in six months.

An object that is further away from the Earth will have a smaller parallax angle than a closer object.

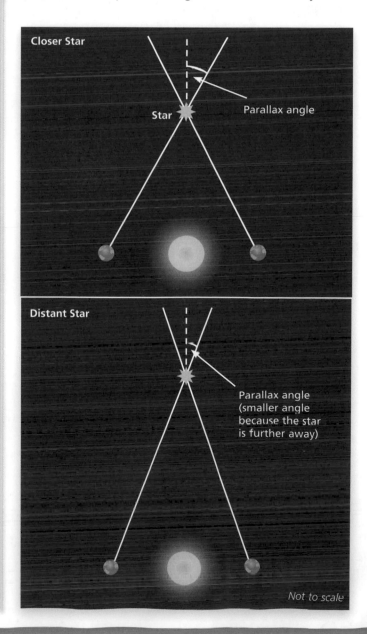

Using Parallax

Astronomers use parallax to measure interstellar distances within our galaxy using the unit **parsec** (**pc**). A parsec is of a similar magnitude to a light-year with 1 parsec equalling roughly $3\frac{1}{4}$ light-years.

Angles are measured in degrees, minutes and seconds. A star that is one parsec away has a parallax angle of one second of an arc.

The distance in parsecs can be found by dividing 1 by the parallax angle, as shown in the following formula:

$$\text{Distance (parsecs)} = \frac{1}{\text{Parallax angle (arcseconds)}}$$

Example

The nearest star to Earth, apart from the Sun, is Proxima Centauri. Astronomers found that it has a parallax angle of 0.77 arcseconds. Calculate its distance from the Earth.

$$\text{Distance} = \frac{1}{\text{Parallax angle}}$$

$$= \frac{1}{0.77 \text{ arcseconds}}$$

$$= \textbf{1.3 parsecs}$$

Parallax is useful for measuring the distance of relatively close objects. For example, the typical interstellar distance is a few parsecs.

Astronomers use the **megaparsec** (**Mpc**) to describe much bigger intergalactic distances, even though these objects are so far away that the parallax angle is too small to measure. For example, the nearest major galaxy, Andromeda, is 770 000 parsecs (0.77Mpc) away.

Measuring Distance Using Brightness

Another method that astronomers use to measure the distance to stars is to observe how **bright** the stars are. This method sounds very simple, i.e. a close star will **appear brighter** than a more distant star.

Unfortunately, not all stars have the same **luminosity**. The luminosity is how much energy the star is emitting, and it depends on the star's size and temperature.

A larger or hotter star will be more luminous than a smaller or cooler star, so it may **appear brighter** even though it is **further away**.

Observed Intensity

The star Antares is 500 light-years from the Earth. There are over 100 000 stars nearer to Earth than Antares, but Antares has luminosity 10 000 times greater than that of the Sun and is the 15th brightest star visible from Earth.

The **observed intensity** of a star depends on its luminosity **and** its distance from the Earth.

A star with a very high luminosity may appear dim if it is very far away.

For example, Sirius appears as the brightest star in the night sky. It is relatively close to the Earth and has a luminosity 23 times greater than the Sun.

A **Cepheid variable** star does not have a constant luminosity. It **pulses** and its luminosity depends on the period of the pulses. The period is equal to 1 ÷ frequency.

HT By measuring the frequency of the pulses of a Cepheid variable star, astronomers can estimate its luminosity. If we know how bright the star **really is** and can see how bright **it appears**, we can work out how **far away it is**.

The Curtis–Shapley Debate

In 1920, a great debate about the scale of the Universe took place between two prominent astronomers – Heber Curtis and Harlow Shapley.

Telescopes had revealed that the Milky Way contained lots of stars and this observation led to the realisation that the Sun was a star in the Milky Way galaxy. Telescopes had also revealed many fuzzy objects in the night sky. These objects were originally called **nebulae** and they played a major role in the debate.

Curtis believed that the Universe consisted of many galaxies like our own, and the fuzzy objects were **distant galaxies**. Shapley believed that the Universe contained only one big galaxy and the nebulae were nearby gas clouds **within the Milky Way**.

In the mid-1920s, Edwin Hubble observed Cepheid variables in one nebula and found that the nebula was much further away than any star in the Milky Way. This observation provided the evidence that the observed nebula was a separate galaxy, supporting Curtis' idea that the Universe contained many different galaxies.

Observations of many Cepheid variables have shown that most nebulae are distant galaxies and have allowed astronomers to measure the distance to these galaxies, and hence determine the scale of the Universe.

The Hubble Constant

By observing Cepheid variable stars in distant galaxies, Edwin Hubble discovered that the Universe was expanding, i.e. the further away a star was, the faster it was moving away. The speed of recession can be calculated by using the Hubble formula.

Speed of recession (km/s)	=	Hubble constant (s^{-1})	×	Distance (km)
Speed of recession (km/s)	=	Hubble constant (km/s per Mpc)	×	Distance (Mpc)

Example 1

A galaxy is a distance of 3×10^{20} km from Earth. If the Hubble constant is $2.33 \times 10^{-18} s^{-1}$, calculate the speed of recession.

Speed of recession = Hubble constant × Distance

$$= (2.33 \times 10^{-18}) \times (3 \times 10^{20})$$

$$= \textbf{700km/s}$$

Example 2

A galaxy is a distance of 10 megaparsecs from Earth. If the Hubble constant is 70km/s per Mpc, calculate the speed of recession.

Speed of recession = 70km/s per Mpc × 10Mpc

$$= \textbf{700km/s}$$

N.B. The speed of recession is the same in both examples. Example 1 uses distance in km and the Hubble constant in s^{-1}. Example 2 uses the astronomical unit of megaparsecs.

HT Example 3

The recession speed of a galaxy is 1200km/s. If the Hubble constant is 70km/s per Mpc, how far from Earth is the galaxy?

$$\text{Distance} = \frac{\text{Speed of recession}}{\text{Hubble constant}}$$

$$= \frac{1200}{70}$$

$$= \textbf{17.1Mpc}$$

Cepheid variable stars are used to accurately calculate the Hubble constant because we know how far away they are. So we can use **redshift** to find out how fast they are moving away (their speed of recession)

This evidence suggests that the whole Universe is expanding and that it might have started around 14 thousand million years ago, from one point, with a huge explosion, known as the **Big Bang**.

HT This effect is exaggerated in galaxies that are further away, which means that the further away a galaxy is, the faster it is moving away from us. This suggests that space itself is expanding.

Astronomers can now use the Hubble constant and redshift data to calculate the distance to other galaxies.

Pressure, Volume and Temperature

Gas pressure is caused by particles in a gas moving about. When a particle collides with an object it exerts a **force**. This force is felt as pressure.

The amount of pressure depends on:
- the number of collisions per second
- the momentum of the particles.

The diagram below illustrates a piston with a gas inside.

As the volume is reduced, the particles have less room to move about and so they collide with the piston more often, increasing the pressure.

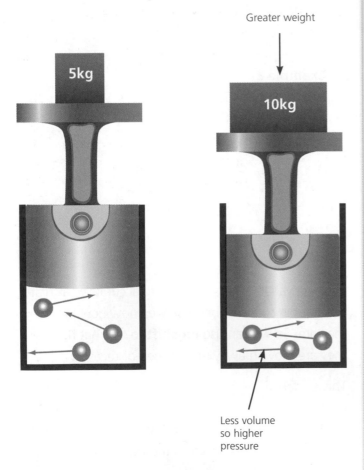

Greater weight

Less volume so higher pressure

If the gas is heated up, the particles will move around faster. This increases their momentum and the force they exert when they collide with the piston. This could have two effects, as shown in the next column.

1 Push the piston back up (increasing the volume):

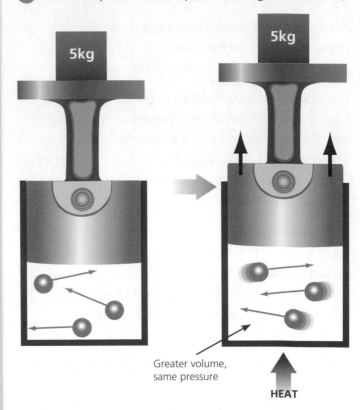

Greater volume, same pressure

HEAT

2 Cause the pressure to increase (if the volume is kept fixed):

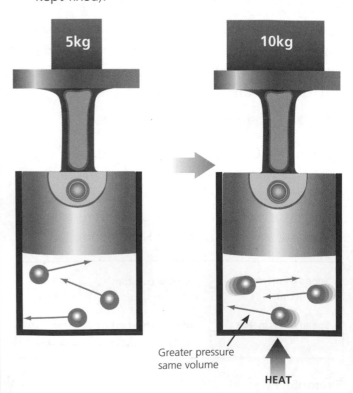

Greater pressure same volume

HEAT

This effect also works in reverse. So, compressing a gas will cause it to increase in temperature. This is often noticed when using a bicycle pump.

Absolute Zero

As the temperature of a gas is reduced, the particles in the gas move more slowly and the pressure falls.

The particles eventually stop moving altogether. At this point the particles have no more energy to lose and the temperature cannot go any lower. This occurs at -273°C, otherwise known as **absolute zero**.

Absolute temperature is a measure of temperature starting at absolute zero and is measured in **Kelvin** (**K**):

- To convert from Kelvin into degrees Celsius, subtract 273.
- To convert from degrees Celsius into Kelvin, add 273.

Example

(a) Convert 400K to °C.

T(°C) = 400 − 273 = **127°C**

(b) Convert 100°C to K.

T(K) = 100 + 273 = **373K**

Pressure Laws

There are three pressure laws that link the pressure, volume and temperature of a fixed mass of gas in a container. They are as follows:

$$\text{Pressure} \times \text{Volume} = \text{Constant}$$

This means if you reduce the volume of a gas, then it will increase the pressure of the gas hitting the sides of the container.

$$\frac{\text{Pressure}}{\text{Temperature}} = \text{Constant}$$

This means if you increase the temperature of a gas inside a container, then it will have more energy and therefore increase the pressure that the gas exerts.

$$\frac{\text{Volume}}{\text{Temperature}} = \text{Constant}$$

This means if you increase the temperature of a gas, then the gas will have more energy and the volume of the gas will increase.

The Structure of a Star

A star has three main parts:

Structure of a Star

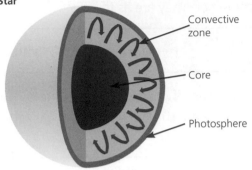

- Convective zone
- Core
- Photosphere

The **core** is the hottest part of the star where fusion takes place. The **convective zone** is where energy is transported to the surface by convection currents (and by photons of radiation). The **photosphere** is where energy is radiated into space.

Like all hot objects, stars emit a continuous range of electromagnetic radiation.

Hotter objects emit radiation of a:
- **higher temperature**
- **higher peak frequency** (i.e. frequency where most energy is emitted) than colder objects.

An object that is red hot emits most of its energy in the red frequency range. The frequency of light given off from a star provides evidence of how hot it is. To analyse the light from stars, we need to look at the frequencies separately.

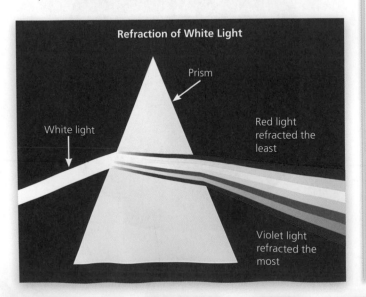

Refraction of White Light

Prism

White light

Red light refracted the least

Violet light refracted the most

The visible spectrum is produced because white light is made up of many different colours. The colours are refracted by different amounts as they pass through a prism – red light is refracted the least and violet is refracted the most. This is because the different colours have different frequencies and, therefore, different wavelengths.

Using a Star's Spectra

The **removal** of an electron from an atom is called **ionisation**.

> **HT** The **movement** of electrons within the atom causes it to emit radiation of specific frequencies called **line spectra**. Different elements have characteristic line spectra. This can be seen by placing a prism at the end of a telescope.

Due to its high temperature, the spectrum from a star is continuous apart from the spectral lines of the elements it contains (these lines are missing because these frequencies are absorbed).

By comparing a star's spectrum to emission spectra from elements, we can find which chemical elements the star contains.

For example, the diagram below compares the emission spectrum for hydrogen and the absorption spectrum that would be seen from the Sun.

The Sun's spectrum is complex, indicating that it contains more than one element. However, by comparing the spectra we can see that the Sun contains hydrogen as well as another element. In the Sun's case we know that this other element is helium.

Hydrogen Spectrum

The Sun's Spectrum

The Beginning of a Star

Stars begin as clouds of gas (mainly hydrogen). As gravity brings these gas clouds together, they become denser. The force of gravity pulls the gas inwards, causing the pressure and temperature to increase. As more gas is drawn in, the force of gravity increases. This compresses the gas so that it becomes hotter and denser and forms a **protostar**.

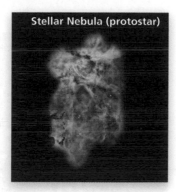

Stellar Nebula (protostar)

Nuclear Fusion

At the beginning of the 20th century, discoveries about the nature of the atom and nuclear processes began to help answer the mystery of where the Sun's energy comes from.

In 1911, there was a ground-breaking experiment – the Rutherford–Geiger–Marsden alpha particle scattering experiment. In this experiment, a thin gold foil was bombarded with alpha particles. The effect on the alpha particles was recorded, and these observations provided the evidence for our current understanding of atoms.

Most alpha particles were seen to **pass straight through** the gold foil. This would indicate that gold atoms were composed of large amounts of open space.

However, some particles were **deflected** slightly and a few were even **deflected back** towards the source. This would indicate that the alpha particles passed close to something positively charged within the atom and were repelled by it.

These observations brought Rutherford and Marsden to conclude that:

- gold atoms, and therefore all atoms, consist largely of empty space with a small, dense core. They called this core the **nucleus**

- the nucleus is positively charged
- the electrons are arranged around the nucleus with a great deal of space between them.

In a star, the temperature and pressure become so high that the hydrogen nuclei fuse into helium nuclei. This is known as **nuclear fusion** and is how the star generates its energy.

Albert Einstein

When scientists first started to look at a star's spectra, they found that through nuclear fusion four hydrogen nuclei were fusing into one helium nuclei and releasing vast amounts of energy. Once they were able to measure the mass of nuclei they found something strange. When you add up the mass of four hydrogen nuclei it does not equal the mass of one helium nucleus. The four hydrogen nuclei are heavier – some of the mass is lost in the formation of the helium nucleus.

> **HT** It was Albert Einstein who came up with the famous equation to explain this:
>
> $$E = mc^2$$
>
> E = The energy produced (J)
> m = The mass lost (kg)
> c = The speed of light (in a vacuum) 300 000 000m/s

The actual mass that is lost is very small, but it produces a vast amount of energy in the fusion of elements up to iron.

Nuclear Reactions in Fusion

Although four hydrogen nuclei fuse together to form one helium nuclei in stars, the reaction takes place over a number of steps. The first step is shown below:

$$^1H + {}^1H = {}^2D + e^+ + Energy$$

D is an isotope of hydrogen called deuterium. The e^+ is called a positron. It is exactly the same as an electron except that it is positively charged and is produced in order to conserve charge. A positron is an example of anti-matter. When anti-matter comes into contact with normal matter it annihilates it, producing a lot of energy.

The Life and Death of a Star

When a star starts nuclear fusion reactions, it is called a **main sequence star** and it is stable. Hydrogen is being converted via fusion into helium.

Towards the end of a star's life, its 'fuel' begins to run out (there is insufficient hydrogen remaining in the core for fusion to continue). The star then undergoes several changes, depending on its size.

When the core hydrogen has been depleted, the star's photosphere becomes cooler. Small stars like the Sun become **red giants**, while larger stars become **red supergiants**.

Red giants and red supergiants continue to release energy by fusing helium into larger nuclei such as carbon, nitrogen and oxygen.

Once the helium has been used up, red giants do not have enough mass to compress the core and continue fusion, so they shrink into hot **white dwarfs** that gradually cool.

Red supergiants have a much greater mass and higher core pressures, so fusion continues to produce larger nuclei such as iron.

Once the core is mostly iron, the star explodes in a supernova, leaving behind a dense **neutron star** or a **black hole**.

Astronomers use a Hertzsprung–Russell diagram to plot the lifetime of a star. If a graph of luminosity (how bright the star is) is plotted against its surface temperature, the following is obtained:

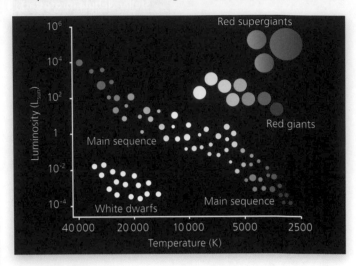

From the graph we can identify regions where supergiants, giants, main sequence and white dwarf stars are located. We can also trace the life of a star from its beginning, to joining the main sequence, and to its death.

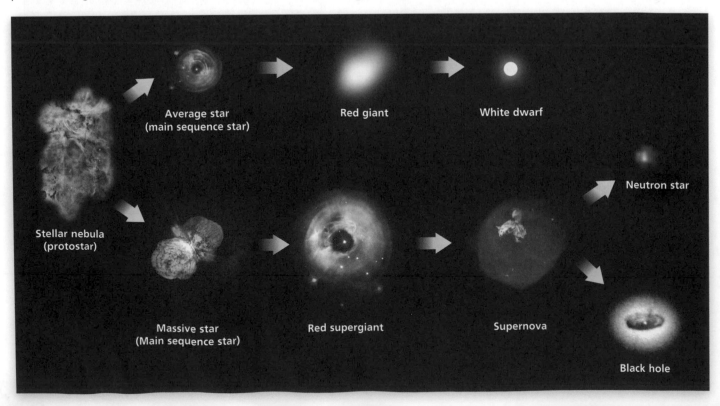

Exoplanets

Scientists have wondered for years whether there are other planets outside our solar system. These are called **exoplanets**. Only recently have astronomers been able to provide evidence for the existence of exoplanets orbiting around their host star.

Detecting an Exoplanet

The easiest way to detect an exoplanet would be to look through an optical telescope, but because the host star is millions of times brighter than the planet, the light reflected from the planet's surface would be drowned out by the star's light. Only planets that are very far away from the star could be seen. In 2004, astronomers using the **Hubble Space Telescope** (see page 80) discovered a huge planet orbiting a star called Formalhaut. It was only discovered due to its vast distance from the host star.

In order to detect planets much closer to the host star, astronomers have had to come up with ingenious methods that have stretched the limits of scientific discovery.

One such way is called the **Radial Velocity** (**RV**) **method**. This relies on the Doppler Effect and the shift in the spectral lines of the star. This Doppler shift happens due to the planet exerting a gravitational force on the star, which causes it to 'wobble'. The planet's size and distance from the star can then be calculated. Most exoplanets have been discovered this way.

Another way to detect planets close to the host star is to measure the amount of light that gets blocked out from the host star when the planet moves in front of the star. This is called the **transit method**. The next generation of space telescopes, such as **Kepler**, have found many exoplanets this way. One such planet recently discovered was similar in size to the Earth.

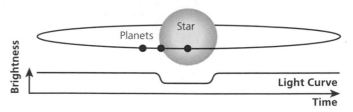

How the Brightness of a Star Changes as Planets Move in Front of It

The Search for Extraterrestrial Intelligence (SETI)

Before it was possible to look for exoplanets, scientists monitored certain parts of the electromagnetic spectrum (e.g. radio waves) to see if extraterrestrial life was trying to contact us. As television signals are high frequency radio waves, we have been sending signals into outer space since the early 1950s. Travelling at the speed of light, our television signals would have reached our neighbouring stars up to a distance of around 50–60 light-years. However, after searching for nearly 40 years, no signals from alien life-forms have been detected.

Ground-based Telescopes

The major optical and infrared astronomical observatories on Earth are mostly situated in Chile, Hawaii, Australia and the Canary Islands. The largest optical telescopes in the world are the 10m aperture reflecting Keck Telescopes at the **Mauna Kea Observatories**, Hawaii.

Hawaii has proven an ideal location for ground-based telescopes for several astronomical reasons:
- Its high altitude means that there is less atmosphere above it to absorb the light from distant objects.
- The lack of nearby cities means that there is less pollution (light and standard) to interfere with the received signal and the air is drier.
- Its equatorial location gives it the best view of solar eclipses.

There are other things that must be considered when deciding where to build an observatory. For example:
- cost
- environmental and social impact near the observatory
- working conditions for employees.

Space-based Telescopes

The most famous space telescope is the Hubble Space Telescope, which was launched in 1990 and designed in collaboration with the European Space Agency and NASA. Its original launch date was delayed by two years because the explosion of the space shuttle *Challenger* shortly after launch resulted in the shuttle fleet being grounded for two years.

After launch it was found that there was a fault on the mirror that required expensive repairs.

Despite these problems, the Hubble telescope has been a great success and provided images of the Universe that could not have been obtained any other way.

The advantages and disadvantages of space telescopes are summarised below.

Advantages
- Avoids the absorption and refraction effects of the Earth's atmosphere.
- Can use parts of the electromagnetic spectrum that the atmosphere absorbs.

Disadvantages
- Very expensive to set up, maintain and repair.
- Risk of harm to astronauts from solar radiation and risk of deadly accidents to astronauts while working in space.
- Uncertainties of the space programme, e.g. launch delays.

Other Telescopes

Radio telescopes use a metal reflector to reflect radio waves onto a receiver. Radio waves are not blocked by clouds so radio telescopes can be sited on the ground. They can detect objects that are too cool to emit much visible or infrared light. However, they have to be much larger than optical telescopes and the images produced are not as clear.

Infrared telescopes work much like optical telescopes. They have a better resolution than radio telescopes and can observe objects too cool to give off visible light. However, infrared light is easily absorbed by the Earth's atmosphere, so this type of telescope needs to be built at high altitude or be space-based.

Funding Developments in Science

Most of the big new telescopes are developed through international collaboration. There are several advantages to this kind of joint venture:
- The cost of manufacturing the telescopes can be shared.
- Astronomers from around the world can book time on telescopes in different countries, allowing them to see the stars from other parts of the Earth.

The telescopes can be accessed directly at the site. They can also be accessed through remote computer control, which can be an advantage because astronomers do not have to travel to each telescope to be able to use it. They can also use the telescopes at convenient times.

Schools in the UK can access the Royal Observatory over the Internet.

This kind of sharing of cost and expertise is essential for many of the big expensive science projects. For example, the Gemini Observatory in Chile, which opened in 2002, was the result of collaboration between Australia and six other countries. The International Space Station has also been jointly funded by the National Aeronautics and Space Administration (NASA), the European Space Agency (ESA), and by space administrations from Japan, Canada and Russia in order to share the cost.

1 The picture below shows a nebula. This is a gas cloud where stars are formed.

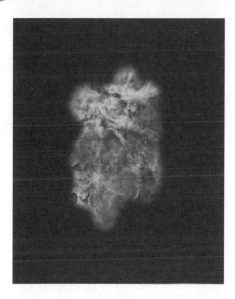

(a) When the nebula collapses to form a protostar, what force causes the star to be formed? **[1]**

(b) When the cloud's temperature gets high enough, what is the name of the process that forms new elements within the new star? **[1]**

(c) What happens to a star after it has used up all its hydrogen and leaves the main sequence? **[1]**

(d) At the end of a star's life, one of two things can happen to it. Explain what would happen to a star similar in size to our Sun. **[2]**

2 A ground-based telescope is shown below.

(a) Explain why most ground-based telescopes are situated on top of mountains at high altitude. **[2]**

(b) Why are ground-based telescopes not situated close to major cities? **[2]**

(c) Other telescopes, such as the Hubble Space Telescope, are put into orbit around the Earth.

   (i) Give two advantages of using space-based telescopes. **[2]**

   (ii) Give two disadvantages of using space-based telescopes. **[2]**

(d) Why can optical telescopes produce very sharp images? **[2]**

Exam Practice Questions

3 Which of the following statements are **true**? Put ticks (✓) in the boxes next to the **three** correct statements. **[3]**

When waves spread out after passing through a narrow gap, it is called diffraction. ☐

The larger the objective lens of a telescope, the more detail the image has. ☐

Cepheid variables pulse with a period related to their mass. ☐

Heber Curtis thought that spiral nebulae were part of the Milky Way. ☐

Scientists believe the Universe began with a 'Big Bang' 14 million years ago. ☐

If the volume of a gas at constant temperature is reduced, the pressure increases. ☐

4 Edwin Hubble linked the speed of recession of a galaxy to its distance from us. Using his equation, work out the speed of recession of a galaxy which is 600Mpc from Earth, if the Hubble constant is 70km/s per Mpc. Show your working and give the units in your answer. **[3]**

5 This question is about nebulae, where stars are born. Complete the following sentences using words from this list. You may use a word more than once. **[4]**

fission	hydrogen	nebula	fusion
dust	helium	oxygen	

The gas in a nebula is mostly .... The force of gravity makes the cloud condense.
As the gas condenses, the temperature of the gas gets higher. When the cloud gets hot enough,
nuclear .... reactions start in which nuclei of .... join together to form
.... and release vast amounts of energy.

HT **6** **(a)** What two factors does the luminosity of a star depend on? **[2]**

(b) What is a Cepheid variable star? **[2]**

(c) Explain how Cepheid variable stars were used to show that the Universe is expanding. **[6]**

✎ *The quality of written communication will be assessed in your answer to this question.*

7 **(a)** What is the ecliptic? **[1]**

(b) Explain why solar eclipses do not occur as often as lunar eclipses. **[2]**

(c) How often do total solar eclipses occur? **[1]**

(d) What is the difference in time between a solar day and a sidereal day? **[1]**

8 The Sun releases energy by nuclear fusion and loses 4×10^9kg of mass every second. Calculate the energy released per second using $E = mc^2$. **[2]**

Answers

Unit A181 (Pages 30–31)

1. (a) | C | A | B | D | F | E | **[1 mark for each up to a maximum of 5.]**
   (b) Five thousand million years
   (c) 300 000km/s
   (d) The distance that light travels in one year.

2. (a) **Radio waves** – Microwaves – **Infrared** – Visible spectrum – **Ultraviolet** – X-rays – **Gamma rays [1 mark for each up to a maximum of 3.]**
   (b) **Any three from:** Sunburn; Skin ageing; Cancer; Kill/Damage cells
   (c) Ultraviolet **[1]**, X-rays **[1]** and gamma rays **[1]** are ionising. Ionising means that the radiation has high enough photon energy to remove an electron from an atom or molecule **[1]**.

3. (a) A resource that will not run out as it can be quickly and easily replaced.
   (b) A resource that cannot be replaced within a lifetime.
   (c) Wind, Wood **and** Tidal water **should be ticked.**
   (d) **This is a model answer which would score full marks:** Burning all fossil fuels releases carbon dioxide, a greenhouse gas, into the atmosphere. The heat released from burning fossil fuels is used to boil water, and water vapour, another greenhouse gas, is released into the atmosphere. Greenhouse gases trap heat from the Sun, causing average global temperatures to rise. This is called global warming. Global warming is likely to cause polar ice caps to melt, raising sea levels and flooding low-lying areas, as well as causing more extreme weather events like hurricanes.
   (e) 2000 × 60 = 120 000J
   [1 mark for correct working but wrong answer.]

4. The movement of the observed wavelength of light towards the red end of the spectrum **[1]**. It tells us that the Universe is expanding **[1]**.

5. (a) Convection currents in the mantle **[1]** cause tectonic plate movement above **[1]**. The tectonic plates try to move towards / past / away from each other, causing a build-up of pressure **[1]**. The sudden release of pressure causes a jolting movement that triggers an earthquake **[1]**.
   (b) **This is a model answer which would score full marks:** The Mid-Atlantic Ridge is a constructive plate boundary. This means that the tectonic plates are moving away from each other. Molten rock rises to the surface, where it solidifies to form new ocean floor. As the rock solidifies, the direction of the Earth's magnetic field is recorded by the rock, as it is magnetised in one direction. The Earth's magnetic field reverses periodically, so there is a striped pattern of rocks either side of the Mid-Atlantic Ridge which is symmetrical. The width of each stripe indicates how long the Earth's magnetic field was orientated in a particular direction.
   (c) A few centimetres

Unit A182 (Pages 61–62)

1. Ionising radiation can break molecules into bits called ions **and** X-rays and gamma rays are ionising radiation **should be ticked**.

2. (a) Momentum = 800kg × 20m/s = 16 000kg m/s **[2 marks for correct answer; 1 mark for correct units.]**
   (b) The momentum will increase.
   (c) The momentum will increase.

3. (a) (i) 0.4A (ii) 0.4A (iii) 3V (iv) 3V
   (b)

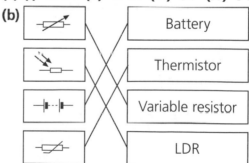

[1 mark for each correct line up to a maximum of 3.]

4. **Any three from:** Medical industry; Nuclear industry; Cosmic rays; X-rays; Food; Radon gas; Stars; Rocks; Uranium; Nuclear weapons testing

5. (Nuclear) fission

6. Between A and B, the bus accelerates from a stationary position **[1]**. The bus then travels at a constant speed to C **[1]**. The bus decelerates between C and D, and is stationary between D and E **[1]**.

Answers

7. **(a)** Mr Benbow and Mr Holmes work for companies that depend on nuclear power.
    (b) Any suitable answer, e.g.
    Mr Benbow – Nuclear energy is not free as the cost of commissioning and decommissioning nuclear plants is very high. Pollution does exist in the form of radioactive waste.
    Miss Manning – The nuclear industry has a better safety record than fossil-fuel power stations.
    Mr Holmes – Renewable energy resources have not been invested in as much as they could. For example, the UK could use wind power to supply a much greater proportion of the electricity needed than it currently does.

8. **This is a model answer which would score full marks:** The vast majority of alpha particles passed through the gold foil without being deflected. Rutherford and Marsden concluded that the atom was mostly empty space. Some alpha particles were deflected through a small angle. Alpha particles are positively charged and they were being repelled by another positively charged object. A very small number of alpha particles were deflected through a large angle (greater than 90°) and for this to happen there must have been a concentration of positive charge in a very small area, i.e. a nucleus of positive charge.

9. $16 \rightarrow 8 \rightarrow 4 \rightarrow 2 \rightarrow 1$ so 4 half-lives have occurred; 4×1 billion = 4 billion years
    [If answer is incorrect, 1 mark for calculation based on half-life.]

Unit A183 (Pages 81–82)

1. **(a)** Gravity
    (b) Nuclear fusion
    (c) It swells to form a red giant or red supergiant.
    (d) When the star leaves the main sequence it forms a red giant **[1]**. After this it shrinks to leave a white dwarf **[1]**.

2. **(a)** There is less distortion from the atmosphere **[1]**, meaning better quality pictures **[1]**.
    (b) There is too much light pollution **[1]**, which interferes with light from space **[1]**.
    (c) (i) Any two from: They avoid absorption **[1]** and refraction caused by the atmosphere **[1]**; They can detect any form of electromagnetic radiation that would have been absorbed by the atmosphere.
    (ii) Any two from: They are very expensive to set up and get into space **[1]**; They are expensive to maintain and repair **[1]**; The uncertainty of the space programme, which can be cancelled **[1]**.
    (d) Light waves have a short wavelength so less diffraction occurs **[1]**. Optical telescopes can be built that have a much larger aperture than the light's wavelength **[1]**.

3. When waves spread out after passing through a narrow gap, it is called diffraction; The larger the objective lens of a telescope, the more detail the image has; **and** If the volume of a gas at constant temperature is reduced, the pressure increases **should be ticked.**

4. Speed of recession = $70 \times 600 = 42\,000$ km/s
    [2 marks for correct answer; 1 mark for correct units.]

5. hydrogen; fusion; hydrogen; helium

6. **(a)** Temperature; Size
    (b) A star that pulses in brightness **[1]**, with a period related to its luminosity **[1]**.
    (c) This is a model answer which would score full marks: Cepheid variable stars in spiral nebulae had their brightness pulses measured and were found to be in separate galaxies. The redshift of the light from these stars indicated that the galaxies were moving away from our own galaxy (the Milky Way). Other Cepheid variable stars were analysed and it was found that the further away a galaxy is, the greater the redshift and the faster the galaxy is moving away. This led to the idea of an expanding Universe, as the Universe was smaller in the past.

7. **(a)** The apparent path the Sun traces out along the sky.
    (b) The Moon does not orbit in the same plane as the Earth **[1]**, so a solar eclipse only happens when the Moon passes through the ecliptic **[1]**.
    (c) Roughly every 18 months
    (d) A sidereal day is four minutes shorter.

8. $E = mc^2 = (4 \times 10^9) \times (9 \times 10^{16}) = 3.6 \times 10^{26}$ J/s (or W)
    [1 for correct working but wrong answer.]

Absolute zero – the lowest temperature possible.

Acceleration – how quickly speed increases or decreases (per second squared).

Activity – the amount of radioactive decay per second.

Alternating current – a current that reverses direction of flow continuously.

Amplitude – the maximum disturbance caused by a wave from the equilibrium position.

Analogue – a signal that varies continuously in amplitude / frequency; can take any value.

Atmosphere – the layer of gas surrounding the Earth.

Atom – the basic unit of a chemical element; an atom contains a nucleus, made of protons and neutrons, which is surrounded by electrons.

Attraction – the force experienced by two objects with mass; the force between two objects with opposite electric charge.

Carrier wave – a wave that carries a signal.

Cell – a component that produces a voltage in a circuit.

Cepheid variable star – a star that has a changing luminosity.

Contamination – contact with or internalisation of radioactive materials, for example by ingesting, inhaling, injecting or being covered in radioactive materials.

Converging lens – a lens in which light rays passing through it are brought to a central point.

Diffraction – the spreading out of a wave as it passes an obstacle and expands into the region beyond the obstacle.

Digital – a signal that uses binary code to represent information; has two states: on (1) and off (0).

Direct current – an electric current that only flows in one direction.

Displacement – the straight line distance between two points. It can be positive or negative.

Distance–time graph – a graph showing distance travelled against time taken; the gradient of the line represents speed.

Electron – a negatively charged subatomic particle that orbits the nucleus of an atom.

Element – a substance that consists of one type of atom.

Focal length – a measure of how strongly an optical system converges (focuses) or diverges light.

Focal point – the point at which rays travelling parallel to the principal axis meet after passing through a converging lens.

Force – a push or pull acting upon an object.

Frequency – the number of times that something happens in a set period of time; the number of times a wave oscillates in one second; measured in hertz.

Global warming – the increase in the average temperature on Earth due to a rise in the level of greenhouse gases in the atmosphere.

Gravitational potential energy – the energy an object has because of its height and mass above the Earth.

Greenhouse effect – the increase in global temperature due to increased levels of greenhouse gases.

Greenhouse gas – a gas in the Earth's atmosphere that absorbs radiation and stops it from leaving the Earth's atmosphere.

Half-life – the time taken for half the radioactive nuclei in a material to decay.

Instantaneous speed – the speed of an object at a particular moment.

Ion – a positively or negatively charged particle formed when an atom, or groups of atoms, loses or gains electrons.

Irradiation – exposure to ionising radiation.

Kinetic energy – the energy possessed by an object because of its movement.

Longitudinal wave – an energy-carrying wave in which the movement of the particles is in line with the direction in which the energy is being transferred.

Luminosity – the brightness of a star, dependent on its size and temperature.

Lunar cycle – describes the Moon's appearance during its 28-day orbit of the Earth.

Lunar eclipse – a point when the Earth's shadow is cast on the Moon.

Momentum – a measure of the state of motion of an object as a product of its mass and velocity.

Neutron – a particle found in the nucleus of an atom that has no electric charge.

Nuclear fusion – light nuclei join to form a heavier nucleus, releasing energy.

Glossary

Parallax – the apparent motion of an object against a background.

Parallel circuit – a circuit in which components are connected in separate loops, so there are two (or more) paths for the current to take.

Parsec – a measurement used for interstellar distances.

Photon – a 'packet' of energy carried by electromagnetic radiation.

Photosynthesis – the chemical process that takes place in green plants where water combines with carbon dioxide to produce glucose using light energy.

Pollutant – a chemical that can harm the environment and health.

Potential difference (voltage) – the difference in electrical energy carried by the charge between two points.

Proton – a positively charged particle found in the nucleus of an atom.

Recycling – to re-use materials that would otherwise be considered as waste.

Redshift – the shift of light towards the red part of the visible spectrum; shows that the Universe is expanding.

Reflection – the change in direction of a wave as it hits a surface.

Refraction – the change in direction and speed of a wave as it passes from one material to another.

Repulsion – a term used to describe the effect when two materials with the same charge push away from (repel) each other.

Resistance – a measure of how hard it is to get a current through a component at a particular potential difference.

Resultant force – the total force acting on an object (the effect of all the forces combined).

Series circuit – a circuit in which there is only one path for the current to take; all components are connected in a single loop.

Signal – any communication that carries a message.

Solar eclipse – a point when the Moon's shadow is cast on the Earth.

Solar system – a collection of stars and planets.

Speed – how far an object travels in a given time; measured in m/s.

Static electricity – the transfer of electrons from one material to another.

Sustainable – capable of being continued with minimal long-term effect on the environment; resources that can be replaced or maintained.

Tectonic plates – huge sections of the Earth's crust that move relative to one another.

Transformer – an electrical device used to change the voltage of alternating currents.

Transverse wave – a wave in which the oscillations (vibrations) are at 90° to the direction of energy transfer.

Universe – billions of galaxies.

Velocity – an object's speed and direction.

Voltage (potential difference) – the difference in electrical energy carried by the charge between two points.

Wavelength – the distance between corresponding points on two adjacent disturbances (waves).

HT **Angular magnification** – a means of measuring magnification based on the angle of rays of light from an object compared to the angle the image appears to make.

Chain reaction – a reaction, for example nuclear fission, that is self-sustaining.

Ecliptic – the apparent path the Sun traces out across the sky.

Isotope – an atom of the same element that contains different numbers of neutrons but the same number of protons.

Nuclear fission – the splitting of atomic nuclei, which is accompanied by a release in energy.

Nuclear reactor – the place where fission occurs in a nuclear power station.

Sidereal day – the time it takes the Earth to rotate 360° about its axis.

Solar day – a full 24 hours.

Subduction – when an oceanic plate is forced under a continental plate.

Velocity–time graph – a graph showing velocity against time taken; the line represents acceleration.

Notes

Notes

Index

Index